W9-CQI-329

Electrostatic Discharge

Understand, Simulate and Fix ESD Problems

Michel Mardiguian

Interference Control Technologies, Inc.
Gainesville, Virginia

Interference Control Technologies, Inc.
Route 625, Gainesville, VA 22065
TEL: (703) 347-0030 FAX: (703) 347-5813

Published 1985
Printed in the United States of America
93 92 91 90 89 88 6 5 4

Library of Congress Catalog Card Number: 85-80686
ISBN: 0-932263-27-5

Acknowledgements

To my wife Christine, who encouraged me with her love, typed most of the manuscript and endured for several months being the spouse of "someone who has a book to write." Many thanks also to Heather Lynskey for her impeccable proofing and to Tim Kingsbury and Shawn Noel for converting my scribbly sketches into fine drawings.

Michel Mardiguian

Preface

Static electricity is the most ancient form of electricity known to man. More than two thousand years ago, the Greeks recognized the attraction between certain materials when they were rubbed together; indeed, the name *electricity* comes from the Greek *elektron*, which means amber. During the 17th and 18th centuries, several key experiments were conducted to understand and measure static electricity. But the discovery of electromagnetism and its formidable breakthrough has rapidly outgrown interest in static electricity. Even today, where the industrial applications of static electricity are not insignificant, they cannot compare with those of electromagnetism and electrodynamics.

Ironically, as much as static electricity was relegated to the attic of scientific evolution, she continuously occupied (I say *she* because *electricity*, in French, is feminine—don't ask me why) the headlines with her undesirable effects. If we consider the thousands of lightning strokes striping the terrestrial atmosphere every minute, we have to realize that our planet and its surrounding clouds are nothing more than a huge electrostatic machine constantly charging and discharging on itself. For decades, people have been learning the hard way that statics can cause explosion of fuels and ammunitions. In 1937, the German flying boat *Hindenburg* arriving in Lakehurst, New Jersey, caught fire while anchoring at its landing mast. What could have been a severe incident became a tragedy: Due to international tension, the USA had put an embargo on helium sales to Germany and the vessel was inflated with hydrogen instead. The resulting fire caused the death of 38 of its hundred or so passengers. Although the causes have not been completely understood, ESD is on top of the list.

More recently, during the 1970's in the USA, a spacecraft launching rocket exploded during the fueling operation, killing three engineers. The cause was, beyond any doubt, identified as ESD. Satellites have paid a heavy toll to ESD, from minor anomalies to severe malfunctions, as in the European Space Agency (ESA) MARECS satellite. In January, 1985, during the assembly of a Pershing missile near Heilbronn, Germany, the motor case, made of Kevlar, was rubbed against the cushioning in its container. ESD caused the four tons of highly flammable propellant to catch fire and the motor exploded, blowing parts 125 meters away, killing three people and injuring nine.

Although such catastrophes are terrible and spectacular, they are quite rare. A more insidious aspect of ESD bloomed in the early 1970's with the arrival of integrated microelectronics. The plants producing integrated circuits started to experience disappointing percentage yields. Once thoroughly investigated, the problem was found to be mostly ESD during all fabrication steps and handling. Though the problem has been fully explained and drastic solutions have been adopted, ESD is still costing millions of dollars a year of pure losses. Considering the astronomical quantities of ICs manufactured every year, the mere fact that 3 to 30% of them die in infancy because of ESD represents an impressive amount of money. To quote G.C. Quinn, Technical Editor of *Electrnoics Test* magazine (April 1984): "The volume range of ESD sensitive components is rising faster than the development and usage of ESD protections . . . Estimating costs of ESD failures not caught at manufacturing inspection is far more difficult. Many of the degraded, *walking wounded* devices may not show up until after termination of the manufacturer's guarantee." Around 1980, Lockheed reported a one-year cost savings of $1.8 million through ESD protection measures that reduced ESD failures by a 16 to 1 ratio. Elementary arithmetic, then, tells us that Lockheed had endured losses of $1.92 million the previous year. Even with severe protection measures, manufacturers still confess that ESD causes 39 to 48% of their IC rejects. The only hope that the plague will ever be dominated is a progressive awareness of people, and the growing use of robots on manufacturing lines.

But the worst was yet to come: With the proliferation of microelectronics in all possible applications, an even bigger number of complaints flourished about erratic errors, transient malfunctions, erased memory, etc. Although the economic losses represented by erroneous transactions and corrupted data of all kinds is difficult to evaluate, it is probably an even bigger figure than the one for chip damage during fabrication. It seems ironic that a physical fact, known for 2500 years for doing nothing but nasty things to us, has continued to defy electronics engineers; solving the problem of transient errors induced by ESD has not been given the same concerted effort as the manufacturing aspect. Most early research was performed by isolated pioneers fighting with their own weapons. Initiating the research themselves, seldom supported by vast budgets, these men used their sagacity and all the resources they could find to investigate a problem for which no measuring techniques existed. They had to invent the tools they needed, and they had to be statisticians, chemists, and RF designers all at once. The names of Ted Madzy, W. Byrne, Michael King, Ralph Calcavecchio, Richard Simonic, and many others that I don't know of, are the names of people to whom all of us who followed are indebted. By mentioning their work, this book will try to render a piece of the recognition they deserve: They paved the road for bringing the understanding of ESD from black magic up to an analytic method. We hope this book will demystify ESD and give a step-by-step strategy for predicting, testing and reducing its effects on electronic equipments.

June 1985 — Michel Mardiguian
Gainesville, Virginia USA

Other Books Published by ICT

1. Carstensen, Russell V., *EMI Control in Boats and Ships,* 1979.
2. Denny, Hugh W., *Grounding for Control of EMI,* 1983.
3. Duff, Dr. William G., *A Handbook on Mobile Communications,* 1980.
4. Duff, Dr. William G. and White, Donald R.J., Volume 5, *Electromagnetic Interference Prediction & Analysis Techniques,* 1972.
5. Feher, Dr. Kamilo, *Digital Modulation Techniques in an Interference Environment,* 1977.
6. Gabrielson, Bruce C., *The Aerospace Engineer's Handbook of Lightning Protection,* 1987.
7. Gard, Michael F., *Electromagnetic Interference Control in Medical Electronics,* 1979.
8. Georgopoulos, Dr. Chris J., *Fiber Optics and Optical Isolators,* 1982.
9. Georgopoulos, Dr. Chris J., *Interference Control in Cable and Device Interfaces,* 1987.
10. Ghose, Rabindra N., *EMP Environment and System Hardness Design,* 1983.
11. Hart, William C. and Malone, Edgar W., *Lightning and Lightning Protection,* 1979.
12. Herman, John R., *Electromagnetic Ambients and Man-Made Noise,* 1979.
13. Hill, James S. and White, Donald R.J., Volume 6, *Electromagnetic Interference Specifications, Standards & Regulations,* 1975.
14. Jansky, Donald M., *Spectrum Management Techniques,* 1977.
15. Mardiguian, Michel, *Interference Control in Computers and Microprocessor-Based Equipment,* 1984.
16. Mardiguian, Michel, *Electrostatic Discharge—Understand, Simulate and Fix ESD Problems,* 1985.
17. Mardiguian, Michel, *How to Control Electrical Noise,* 1983.
18. Smith, Albert A., *Coupling of External Electromagnetic Fields to Transmission Lines,* 1986.
19. White, Donald R.J., *A Handbook on Electromagnetic Shielding Materials and Performance,* 1980.
20. White, Donald R.J., *Electrical Filters—Synthesis, Design & Applications,* 1980.
21. White, Donald R.J., *EMI Control in the Design of Printed Circuit Boards and Backplanes,* 1982. (Also available in French.)
22. White, Donald R.J. and Mardiguian, Michel, *EMI Control Methodology & Procedures,* 1985.
23. White, Donald R.J., Volume 1, *Electrical Noise and EMI Specifications,* 1971.
24. White, Donald R.J., Volume 2, *Electromagnetic Interference Test Methods and Procedures,* 1980.
25. White, Donald, R.J., Volume 3, *Electromagnetic Interference Control Methods & Techniques,* 1973.
26. White, Donald R.J., Volume 4, *Electromagnetic Interference Test Instrumentation Systems,* 1980.
27. Duff, William G., and White, Donald R.J., Volume 5, *Prediction and Analysis Techniques,* 1970.
28. White, Donald R.J., Volume 6, *EMI Specifications, Standards and Regulations,* 1973.
29. White, Donald R.J., *Shielding Design Methodology and Procedures,* 1986.
30. *EMC Technology 1982 Anthology*
31. *EMC EXPO Records 1986, 1987, 1988*

All of the books listed above are available for purchase from Interference Control Technologies, Inc., Don White Consultants, Subsidiary, State Route 625, P.O. Box D, Gainesville, Virginia 22065 USA. Telephone: (703) 347-0030; Telex: 89-9165 DWCI GAIV.

Table of Contents

List of Figures

Figure	Title	Page

List of Tables

Chapter 1

The ESD Phenomenon

Although a thorough description of the electrostatic phenomenon is beyond the scope of this handbook and has been covered by several authors (Refs. 1,1a,2,3), it might prove useful to review briefly how static electricity takes place, what the contributing parameters are and why, ultimately, it results abruptly in its threatening consequence: electrostatic discharge (ESD).

1.1 Physics Involved

Any material is made of atoms. Unless submitted to certain external influences (heating, rubbing, electrical stress, etc.) the atom is in equilibrium; that is, the amount of negative charges represented by the electrons orbiting around the nucleus is exactly balanced by an equal number of positive charges, or protons, aggregated in the nucleus. Therefore, the net electric charge seen from the ouside is zero.

In metals, the mobility of electrons is such that the conditions of equilibrium will always exist; i.e., no significant static field will exist between different zones of the same piece of metal. With non-conductive materials, however, the lesser mobility of electrons does not provide such rapid recombination of charge unbalance. If heated or rubbed strongly (which also creates heat), a non-conductor will free up electrons.

Depending on the nature of its outer valence orbit, a non-conductive material may be likely to give up electrons or to capture wandering electrons. A non-conductive material which gives up an electron as in Fig. 1.1 will become positively charged. Such unbalanced atoms are called positive ions.

A non-conductive material which takes extra electrons will become negatively charged and its atoms with an excess of electrons are called negative ions. Charges with like sign repel while charges with opposite signs attract. Therefore, it seems that nature will rapidly take care of the unbalance by recombining the charges. Unfortunately, while this recombination is instantaneous in metals (this is, indeed, how a current flows), the high resistance of non-conductive materials makes it unlikely to happen until such a high gradient of field is reached that either an arc or a mechanical attraction will occur. Besides rubbing or heating, which is the actual generation mechanism, an object can become charged through *contact* with another previously charged object.

The ability of non-conductive materials to acquire electrostatic

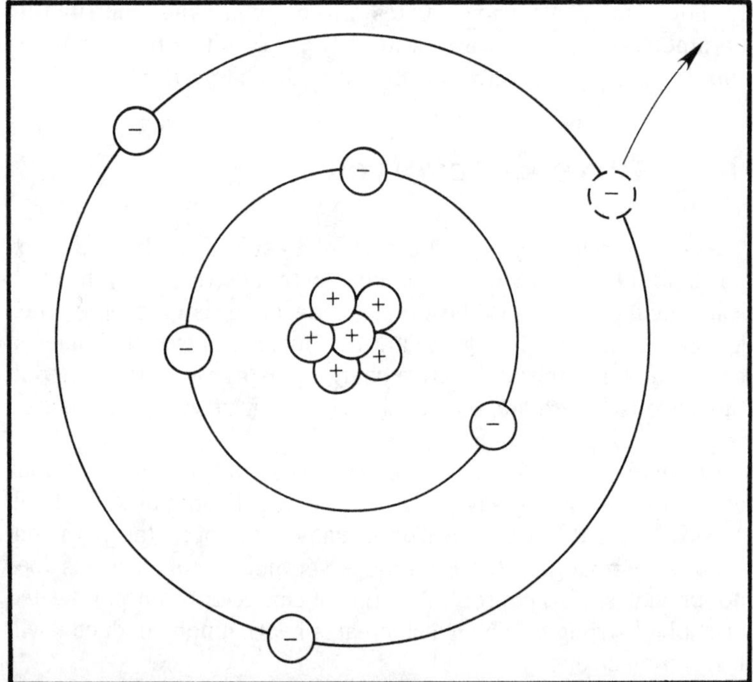

Figure 1.1—If, due to heating, rubbing, etc., one electron leaves the orbit, the material is left with 6 protons and only 5 electrons.

charges is frequently shown as a triboelectric scale such as the one shown in Fig. 1.2. The materials labeled "positive" will take on a positive charge every time they come into contact with a material lower on the scale.

Although this kind of scale is true overall, the precise ranking of each material within the scale should not be relied upon in real life situations (A. Testone, in his excellent booklet in Ref. 2, has

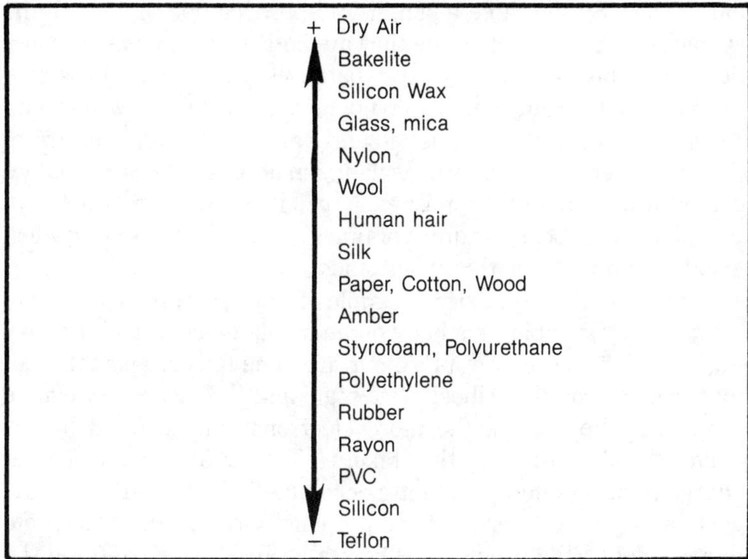

Figure 1.2—Triboelectric series

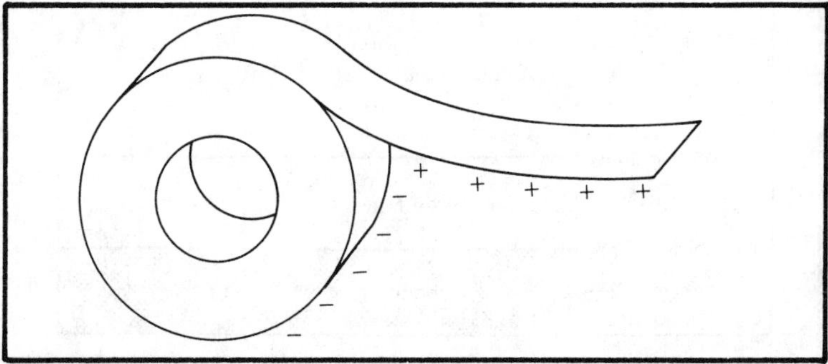

Figure 1.3—A strip of acetate tape is attracted to the surface of the tape reel. Two insulating materials of the same nature can develop opposite charges if enough friction is applied.

shown how deceptive such triboelectric tables can be.) For example, let's take a reel of ordinary office scotch tape. If we quickly unwind a length of tape, everyone knows that this segment becomes charged and can attract small particles of dust, hairs, etc. But, according to the table, this piece of acetate cannot develop an electric field against itself. Indeed, if we cut a piece of this section of tape and approach it from the loose end, the two pieces of tape repel each other. By the same token, if we approach one piece with the surface of the reel itself, since they are both acetate, no field should develop. This is similar to two batteries placed side by side—terminals of the same polarity do not create any field between them. In fact, however, the tape is violently attracted to the surface of the reel. Therefore, even two insulating materials of the same nature can eventually develop opposite charges if sufficient friction is applied.* This happens hundreds of times a day in a photocopier when a foil is slipped over the paper stack.

Now consider the classical example of a person walking on a synthetic carpet, rubbing his body on an insulated chair pad, or moving his nylon sleeve over a PVC surface: The farther apart the two materials are on the triboelectric scale and the faster the relative motion of the person, the more electrons will be freed by the "givers" and captured by the "takers." This creates an unbalanced charge and, therefore, a latent electric field. Testone (Ref. 2) suggests a "figure of merit" of the propensity of materials to create more or less ESD problems. The scale (Fig. 1.4) is based on the

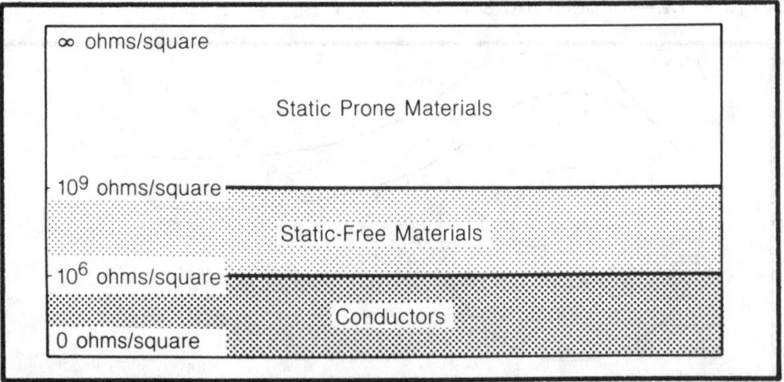

Figure 1.4—Propensity of materials to create ESD problems based on surface resistance in ohms/square

*A counter-argument to this would be that actually *there is* another material involved—the layer of adhesive on the tape. But the experience can be repeated with a reel of wrapping plastic, free of any additional coating, showing the same results.

surface resistance in ohms per square (i.e., the resistance of a sample of square size, whether it is 1 cm² or 1 m², yields the same results). Materials with more than 10^9 ohms/square are likely to develop electrostatic potentials which will not bleed off by themselves due to the high insulation of the material. Materials with less than 10^9 ohms/square, even if not real conductors, will not keep the charge imbalance very long because recombination will occur through the material itself.

Aguet (Ref. 3) relates the propensity to electrostatic discharge to the dielectric constant of the materials. He indicates the surface charge density ϱs:

$$\varrho s = 15\times10^{-6}\ (\epsilon r_1 - \epsilon r_2)\ \text{coulomb/m}^2 \tag{1.1}$$

where ϵr_1 and ϵr_2 are the relative permittivity of the two materials. For instance, if one looks at a rubber shoe sole ($\epsilon r \cong 2.5$) representing 250 cm² and a nylon carpet ($\epsilon r \cong 5$), the maximum total charge can be:

$$\begin{aligned} Q &= 15\times10^{-6}\ (5-2.5)\times25\times10^{-4}\ \text{m}^2 \\ &= 0.93\times10^{-6}\ \text{coulombs} \end{aligned}$$

If the corresponding foot-to-ground capacitance is about 100 pF, the static voltage is given by:

$$V = \frac{Q}{C} = \frac{0.93\times10^{-6}\ \text{coulomb}}{10^{-10}\ \text{farad}} = 9{,}300\ \text{volts} \tag{1.2}$$

It might seem, then, that there is practically no upper limit to what voltage a person can attain. Why not 50 kV or 100 kV? Richman, in his very illustrative pamphlet on ESD (Ref. 4), explains that personnel electrostatic voltage cannot exceed 35 kV even in the most extreme case because:

- human body capacitance, no matter what we do, cannot drop below 30-40 pF, a value that Richman calls our "capacitance to infinity."
- above 35 kV, corona will self-limit our voltage by bleeding off charge; i.e., the assumption of constant Q no longer applies.

1.2 Influencing Parameters

Once the type of material present is known, the most important parameter is relative humidity. It is well known that, during the winter and spring season, all integrated circuit manufacturers have recorded an increased rate of chip failures, all field engineers report an increased number of service calls for computer failures, etc. Several things happen when relative humidity is low:

- Normally, the moisture content in the air tends to lower the surface resistance of floors, carpets, table mats, etc. by letting wet particles create a vaguely conductive (or less than 10^{-9} ohms/square) film over an otherwise insulating surface. If the relative humidity decreases, this favorable phenomenon disappears.
- The air itself, being dry, becomes a part of the electrostatic build-up mechanism every time there is an air flow (wind, air-conditioning, blower) passing over an insulated surface.

Many evaluations have been made of the electrostatic voltages reached by a person walking on several types of floors. Generally these tests are made using a kind of "standard walking procedure." The person walks a given number of steps wearing a certain type of shoe, then his or her charging voltage is immediately measured with an electrostatic voltmeter having a quasi-infinite input impedance. The importance of measuring the voltage immediately, and preferably having the same short time for all experiments, is shown on Fig. 1.5. If the time elapsed between the end of the voltage build-up phase and the instant of the measurement is not kept constant, comparisons between materials, clothes, shoes, etc. become inaccurate.

Fig. 1.6, from Ref. 6, shows the magnitudes range of generated electrostatic voltages for several floor types and two values of the relative humidity (RH). On the left side, the voltages are shown for a relative humidity of 50%.

Even for a notoriously bad type of carpet like nylon, the voltages stay within 1-3 kV. Note that this is already enough to kill an integrated circuit if the person handles a module or a PCB directly.

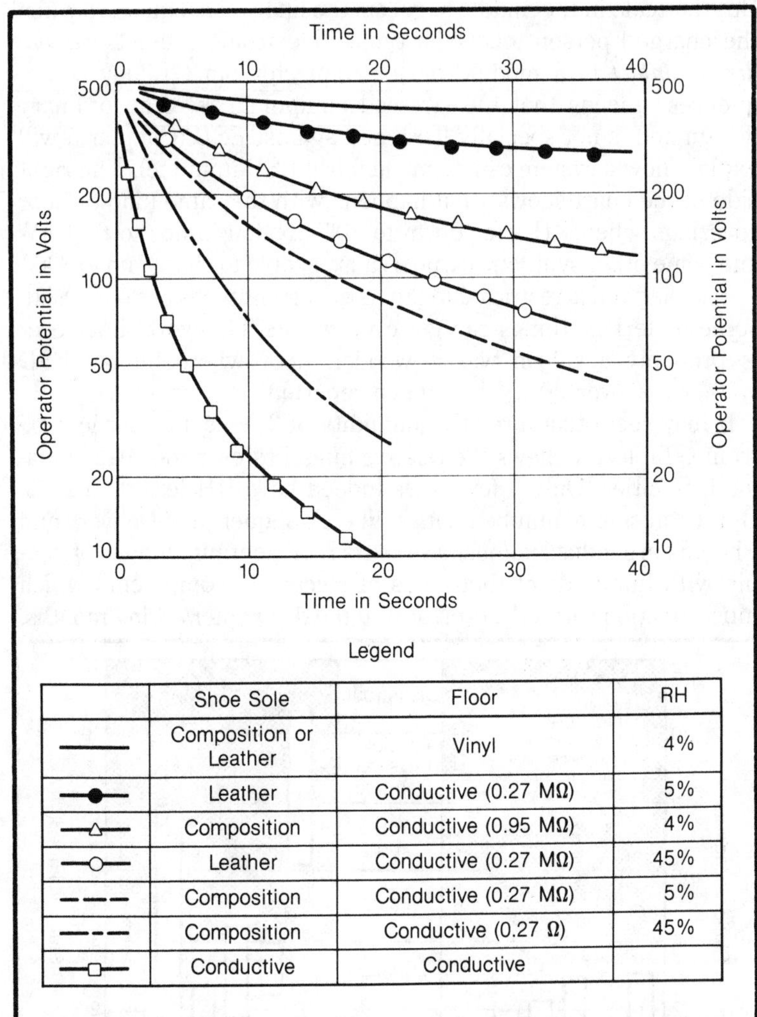

	Shoe Sole	Floor	RH
———	Composition or Leather	Vinyl	4%
—●—	Leather	Conductive (0.27 MΩ)	5%
—△—	Composition	Conductive (0.95 MΩ)	4%
—○—	Leather	Conductive (0.27 MΩ)	45%
— — —	Composition	Conductive (0.27 MΩ)	5%
——— – –	Composition	Conductive (0.27 Ω)	45%
—□—	Conductive	Conductive	

Figure 1.5—Electrostatically generated voltages decay at a rate that is dependent on relative humidity, type of floor covering and type of clothing worn by personnel. Decay times can take several minutes to reach safe levels (from Ref. 5).

But the scale in the middle suggests the likely consequences when the charged person touches a typical electronic cabinet, *without direct contact* to a module pin or connector pin. "Likely consequences" means that the stressed equipment is of an ordinary design and is not specially hardened against ESD (Chapter 5 will explain how a system can be made fairly ESD immune). The right side of the chart shows what happens with the same kind of floor coverings when RH goes down to 20%. Nylon jumps to 6-11 kV, and some other synthetic carpets cause people to charge up to 8 kV.

The diagram is restricted to the most current types of floors. With some materials, things can get even worse. The worst floor ever is probably a silicon waxed wooden floor where human ESD voltages of over 20 kV have been reported.

It may seem that a relative humidity of 20% is a rather low extreme. Table 1.1 shows the relative humidity over the year in major U.S. cities. Only a few cities indeed have RH less than 20%. But in absolute number, cities like Albuquerque, Denver, and Phoenix are industrial/business areas representing millions of people with hundreds of thousands of electronic equipments which must function correctly, even during the dry winter/spring months.

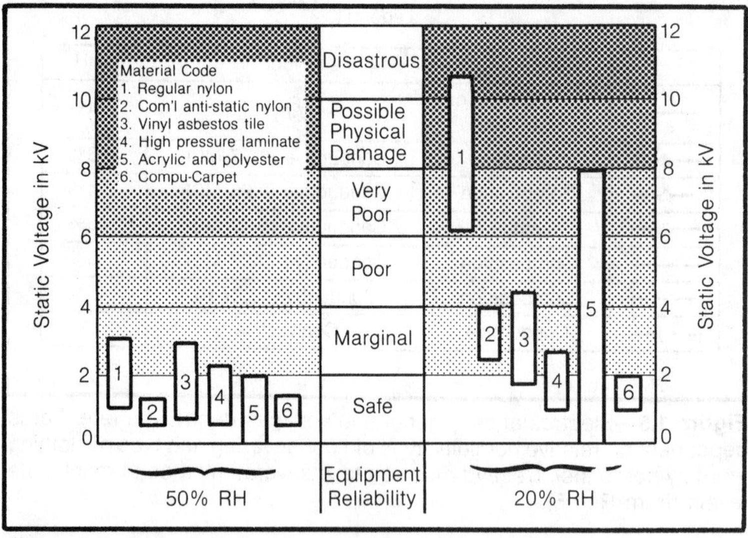

Figure 1.6—Typical static voltages generated by walking on common floor covering materials (see Ref. 6)

Table 1.1—Relative humidity in selected U.S. cities

STATE	STATION	Length of record (yr.)	Jan.		Mar.		May		July		Sept.		Nov.		Annual	
			7:00 a.m.	1:00 p.m.	7:00 a.m.	1:00 p.m.	7:00 a.m.	1:00 p.m.	7:00 a.m.	1:00 p.m.	7:00 a.m.	1:00 p.m.	7:00 a.m.	1:00 p.m.	7:00 a.m.	1:00 p.m.
Ala	Mobile	17	81	62	83	56	86	54	89	61	88	61	85	56	85	57
Alaska	Juneau	36	79	76	79	69	74	62	81	70	87	77	85	81	81	73
Ariz	Phoenix	19	45	32	33	23	18	13	28	20	31	23	37	28	32	22
Ark	Little Rock	19	80	62	78	56	86	57	87	58	89	59	82	57	83	57
Calif	Los Angeles	20	56	60	62	65	66	66	69	68	66	67	57	63	62	65
	Sacramento	19	85	70	68	52	50	36	47	28	50	31	75	59	63	46
	San Francisco	20	79	67	70	63	64	60	66	60	65	59	73	64	69	62
Colo	Denver	19	45	48	42	40	39	37	35	35	38	34	45	50	40	40
Conn	Hartford	20	72	57	72	53	73	47	79	51	87	56	80	57	77	53
Del	Wilmington	32	75	60	74	53	76	53	80	54	85	56	80	56	78	55
D.C	Washington	19	68	54	68	49	72	51	76	53	80	56	74	53	73	52
Fla	Jacksonville	43	87	57	85	49	84	50	88	58	91	62	89	55	87	55
	Miami	15	84	60	82	56	82	61	85	64	89	68	85	61	84	62
Ga	Atlanta	19	78	60	78	51	83	55	90	63	90	61	82	54	83	57
Hawaii	Honolulu	10	80	63	73	58	67	55	66	51	66	52	74	59	71	56
Idaho	Boise	40	73	70	55	44	45	34	33	21	39	30	65	60	52	44
Ill	Chicago	21	76	68	79	62	76	54	81	57	84	57	80	64	79	60
	Peoria	20	77	68	81	64	81	57	85	58	88	58	83	66	82	61
Ind	Indianapolis	20	81	70	80	63	82	57	87	60	91	59	85	67	84	62
Iowa	Des Moines	18	76	68	78	63	78	55	81	57	84	59	79	64	80	61
Kans	Wichita	26	79	63	76	54	83	55	78	49	82	55	79	57	79	55
Ky	Louisville	19	77	65	76	59	83	56	86	59	89	60	79	61	81	59
La	New Orleans	31	85	67	85	61	89	60	91	66	89	66	86	61	88	63
Maine	Portland	39	77	62	75	59	75	58	80	60	86	61	84	63	80	60
Md	Baltimore	26	71	58	71	51	77	53	82	53	85	56	77	55	77	54
Mass	Boston	15	68	58	68	57	71	58	73	56	79	61	75	61	72	58
Mich	Detroit	45	78	69	77	60	71	51	74	51	82	55	79	64	77	58
	Sault Ste. Marie	38	82	76	83	68	80	56	89	62	92	67	87	76	85	67
Minn	Duluth	18	74	68	77	63	76	54	83	58	86	63	80	69	80	63
	Minneapolis-St. Paul	20	72	66	76	63	76	52	81	54	86	60	80	66	79	60
Miss	Jackson	16	87	66	88	57	92	56	93	60	94	61	91	58	91	59
Mo	Kansas City	7	74	65	78	62	83	59	82	56	86	60	78	62	80	60
	St. Louis	19	83	66	82	59	83	56	86	57	91	59	85	63	84	60
Mont	Great Falls	18	63	62	53	48	46	40	38	29	45	36	55	54	50	45
Nebr	Omaha	15	76	65	76	57	79	55	83	56	87	61	80	62	80	59
Nev	Reno	16	68	50	47	33	33	25	28	19	34	21	56	41	44	31
N.H	Concord	14	74	60	78	56	79	48	86	52	91	57	85	61	82	55
N.J	Atlantic City	15	75	58	76	55	79	57	84	58	86	59	82	57	80	57
N. Mex	Albuquerque	19	50	39	32	23	24	17	35	27	40	31	42	35	37	28
N.Y	Albany	14	79	64	74	54	76	52	80	54	88	59	81	63	79	57
	Buffalo	19	80	74	80	67	76	56	79	55	83	60	82	70	80	63
	New York	58	68	60	67	55	71	53	75	55	79	57	73	59	72	56
N.C	Charlotte	19	78	56	80	51	84	54	88	59	90	58	84	53	83	54
	Raleigh	15	79	55	80	49	87	56	90	59	93	60	85	51	85	54
N. Dak	Bismarck	20	72	66	78	62	79	49	83	47	82	49	79	62	79	56
Ohio	Cincinnati	17	79	68	77	60	80	54	85	57	88	59	80	63	81	60
	Cleveland	19	77	70	78	65	77	58	82	58	84	61	79	66	79	63
	Columbus	20	76	68	73	58	79	55	84	56	88	59	81	64	80	59
Okla	Oklahoma City	14	79	61	75	52	83	58	81	50	84	56	78	54	80	54
Oreg	Portland	39	82	76	72	60	66	53	61	45	67	49	82	74	72	60
Pa	Philadelphia	20	73	59	71	53	75	53	79	55	83	57	77	56	76	55
	Pittsburgh	19	76	66	74	58	76	51	82	53	86	57	79	63	78	57
R.I	Providence	16	72	58	70	54	72	52	77	57	82	57	78	59	75	55
S.C	Columbia	13	82	55	83	48	87	51	89	56	93	58	88	49	87	52
S. Dak	Sioux Falls	16	74	67	80	63	80	53	81	52	85	57	82	64	80	59
Tenn	Memphis	40	78	63	76	56	82	55	85	57	86	56	79	55	81	57
	Nashville	14	80	65	78	54	87	56	91	59	92	61	81	59	85	58
Tex	Dallas-Fort Worth	16	80	61	80	57	88	61	81	50	87	59	81	56	82	57
	El Paso	19	44	34	28	20	22	15	39	29	44	34	38	32	35	27
	Houston	10	88	66	89	60	94	60	94	59	95	65	91	60	92	60
Utah	Salt Lake City	20	70	68	52	45	37	31	26	20	34	27	57	58	46	42
Vt	Burlington	14	70	64	73	59	75	51	80	54	87	64	79	69	78	60
Va	Norfolk	31	75	59	74	54	78	57	82	60	84	62	79	56	78	58
	Richmond	45	81	57	78	49	79	51	85	57	90	57	85	50	83	53
Wash	Seattle-Tacoma	20	78	74	74	62	67	54	65	48	74	59	80	74	73	62
	Spokane	20	82	77	67	54	52	40	39	25	50	34	81	74	62	51
W. Va	Charleston	32	77	63	74	53	82	50	90	61	91	56	80	56	82	56
Wis	Milwaukee	19	75	68	79	66	79	61	82	61	87	63	81	67	81	65
Wyo	Cheyenne	20	45	48	44	45	39	41	34	37	35	36	42	47	40	42
P.R	San Juan	24	80	64	77	60	77	65	78	66	79	67	81	66	79	64

Please note: These figures, given in percentages, represent the averages for the period of record through 1979. Eastern Standard Times are indicated.

Courtesy of American Express

In fact, the problem is more critical than Table 1.1 indicates: The recorded relative humidities are those found outdoors by the weather bureau. A significant difference may exist between the outdoor RH and its actual value in a heated building. This is due to the fact that, given a same quantity of water, warm air has a greater ability to absorb moisture; therefore, its relative humidity (compared to saturation) is lower. Fig. 1.7 shows that in a restricted space with temperature T_2, the relative humidity will be:

$$RH_{2(T_2)} = RH_{1(T_1)} \times \frac{T_2}{T_1} e^{c(1/T_2 - 1/T_1)} \tag{1.3}$$

where RH is the relative humidity of the outer ambient at temperature T_1 and C is a constant equal to 5370 between −20°C, and +70°C. T_1 and T_2 are given in °Kelvin.

The equation has been plotted for few typical situations. For in-

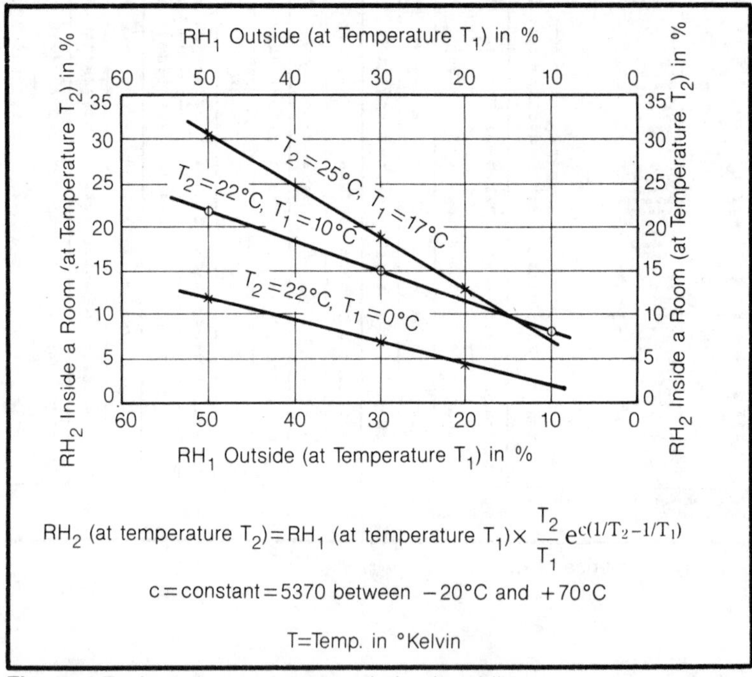

Figure 1.7—Actual vs. apparent relative humidity

stance, on a winter day where the outside temperature is 0°C (273°K) and the RH is about 40%, the actual RH in a room heated to 22°C will be only 9%! Unless the heating or air-conditioning systems compensate for this lack of water vapor, which they generally do rather poorly, or unless a humidifier is installed, the ESD risk is very high.

Besides the type of material and the relative humidity, other factors play a role in the severity of the human electrostatic charge:

- type of clothing
- speed and manner of walking
- sex and size
- body capacitance
- body resistance

The two last contributors will be discussed in the next section because they strongly influence the dynamic characteristics of the discharge.

1.3 Various Types of Electrostatic Charging from Humans and Charged Objects

Although an infinity of ESD cases has been reported, the ones plaguing the electronic industry belong to either the *human body discharge* or the charged *object discharge.* So far electrostatic charges generated by human beings have been emphasized. Figure 1.8 shows some of the classical ways that a human body can generate static charges. Notice that, depending on the nature of the two materials being rubbed together, *the person can exhibit positive or negative charging.*

Although humans have a tendency to treat themselves as a very special kind, physics do not care about this and treat humans as a mere conglomerate of materials, vaguely conductive. There are thousands of occasions where the human body *is not* the electrostatic generator, but is simply the carrier or is not even involved at all. An example of a human as a "carrier" occurs when a person gets out of his car after a ride on a bright, cold, winter day. If he puts his foot on the ground while his hand is touching the

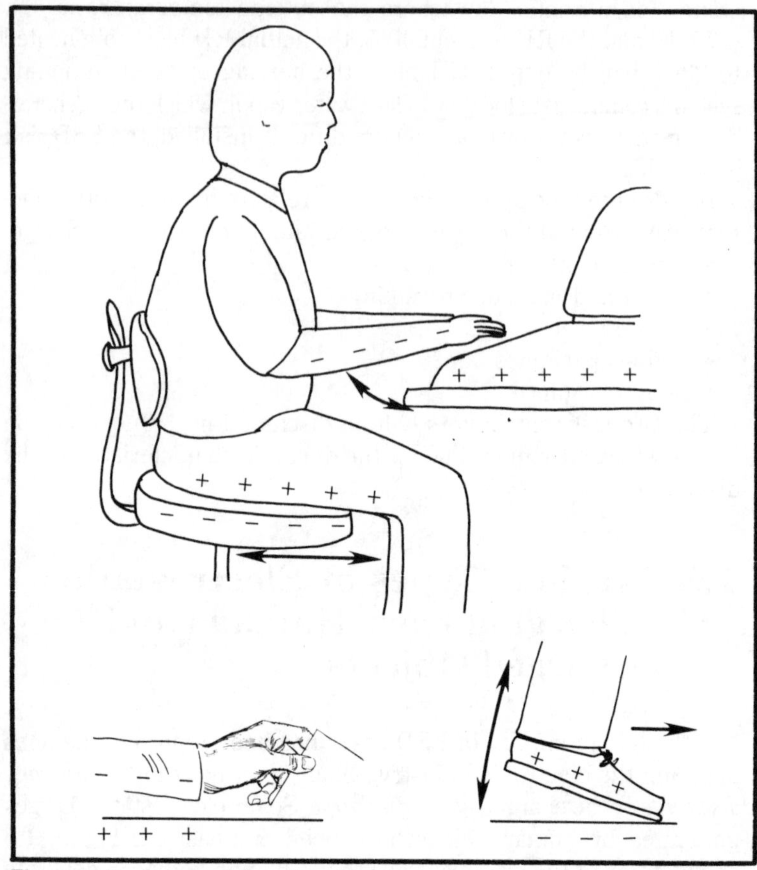

Figure 1.8—Some of the classic ways a human can generate ESD

door handle, he may feel an ESD zap. The body was not the electrostatic generator in this case, the car was. There have been cases reported of toll gate attendants who could not stand their jobs because of too many ESD zaps when drivers handed them the money!

The following is a list of some non-human ESD sources:

- Wheelchairs and rolling furniture
- Rubber or textile belts and conveyors and their pulleys/rollers
- Cooling fans with plastic rotor blades
- Paper movement (printers, copiers)

- Rapid flow or friction of gas, liquid or granules against an insulating material or ungrounded conductor, such as:
  - cleaning with an airgun
  - PVC "skin-packing" with hot air blast
  - cleaning with a solvent
  - fuel lines
  - loading or dumping grain, etc., in silos
  - rocket exhaust nozzle
  - radomes, fiberglass hoods and tips
  - thermal blankets (spacecrafts)

Also, it is important to remember that a non-conductive object can become charged from contact with a previously charged object. This static-contaminated object will, in turn, become a hazard for electronics.

In all cases, whether a person is involved or not, the charged object will "seek" the first opportunity to recombine the unbalanced charges. This may occur smoothly by a progressive bleed of charges through a moderately conductive path, or it may be abruptly and generally accompanied by an arc.

In the case of a "self" recombination, a small amount of current will flow during a certain time, and the result will generally be harmless. In the case of an abrupt recombination, the discharge will occur during a very short time due to the high gradients in-

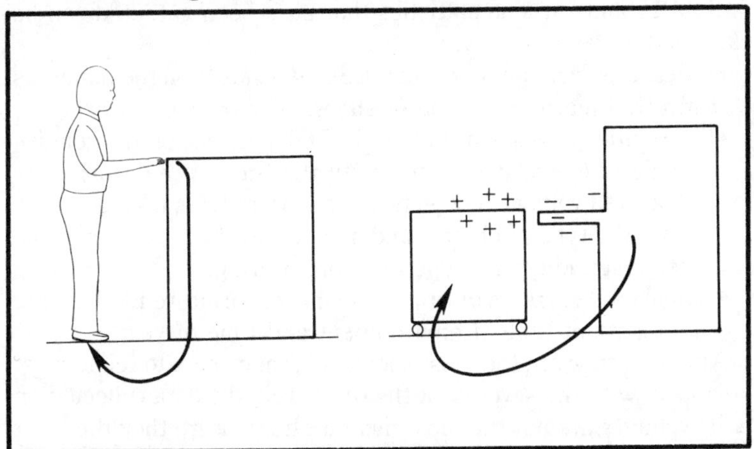

Figure 1.9—ESD with or without a human body involved

volved, and the corresponding current will be high, since the *average* current is:

$$I_{Amp} = \frac{Q_{coulombs}}{t_{seconds}} \tag{1.4}$$

When a discharge of microcoulombs takes place within tens of nanoseconds, the *average* currents amount to several amperes, with peak values that can reach up to a hundred amperes!

1.4 Statistics of Voltages and Currents Reached During ESD

Although the ESD phenomenon has been fought for decades by the electric and electronic industries, it was only during the early 1970's that thorough studies were carried out on its dynamic parameters.

Measurement of voltages and/or currents encountered during real or recreated ESD situations, sampled over a certain period of time or among a certain number of individuals, are reported in Refs. 6, 7, 8, 9, 10 and 11. The statistics that have been gathered can be classified as follows:

- Measured voltages of human ESD dependent on the materials involved (garments, type of shoes, type of floor covering)
- Measured voltages of human ESD dependent on the relative humidity (correlated or not with the time of the year)
- Measured voltage of objects and furniture ESD, taking into account the type of objects and sometimes the type of environment (humidity and type of floor covering)
- Measured currents during personnel or furniture ESD, taking into account type of environment and time of year

Although they are the most spectacular and seem to relate most obviously with the severity of the discharge, the data collected on ESD voltages are not the most meaningful, nor are they the most crucial when trying to develop representative specifications for ESD simulation.

The voltages at which the persons were charged during the measurement campaign can be an ambiguous or inaccurate data base. Were the voltages measured at their peak, right after a static build-up? Or were they measured right at the moment of an actual discharge? What were the mean value and standard deviation of the voltage decay between its peak value (when generated) and its value at the exact moment of the discharge? Were the people in the study aware that ESD voltages were gathered? Every statistician knows that people in such surveys often tend to "sympathize" with the experimenter and to *help him to find what he wants to find* (in our case, for instance, by shuffling their shoes more conscientiously on the carpet!). Statistics is a serious discipline requiring some specific precautions. In some of the often mentioned experiments, it is not known for certain that these precautions were taken, or were feasible.

Drawing a parallel between ESD and lightning, one could say that focusing on ESD voltages only is probably as irrelevant as concentrating on cloud-to-earth voltages when dealing with lightning strokes: All sound statistics on lightning severity are based on lightning currents. Similarly, statistics based on ESD current seem the most dependable and usable.

As for any transient or random event, statistical analysis is important for determining the probability that a certain value of ESD will be reached or exceeded. Do we want to test the immunity of an electronic device or equipment to something which can happen once a day or once a year? How many failures per week or month do we risk by testing a machine only to a certain ESD level?

During a particularly cold and dry spring (early April 1982) in Virginia, the author took a limited survey of the number of discharges (as they were felt by people) in a set of offices where about 12 people were using one copier, a telex, a desk-top computer, a postage meter, and a word processor.

Discharges were recorded on Tuesdays, Wednesdays, and Thursdays for three weeks. The discharges did not necessarily cause equipment malfunction—in fact most of them did not—but the purpose of the survey was to count the ESD *events*, not the eventual failures. It must be also noted that discharges of less than 1500-2000 volts are generally not felt by a person and therefore were not recorded. Such discharges represent no risk to normal electronic equipment, but their absence in a sampling could skew the statistical analysis and produce overly pessimistic conclusions. The results of this survey are shown in Fig. 1.10.

During two campaigns spread over several years, Simonic (Refs. 10 and 11) compiled with impeccable rigor thousands of measurements of both human and furniture ESD in quasi real-life conditions. His work represents such an outstanding contribution to the ESD subject that it deserves a detailed analysis.

1.4.1 Personnel ESD Statistics

The first compilation made by Simonic covered personnel ESD events. The survey was run for 16 months and its analysis allows for predicting, given a human contact discharge, the probability of reaching a peak discharge current I.

The highlights of the analysis are the following:

- The room surveyed had a high human activity (terminal room with 16 operator-attended stations).
- The room had uncontrolled (or poorly controlled) RH and a wool carpet.

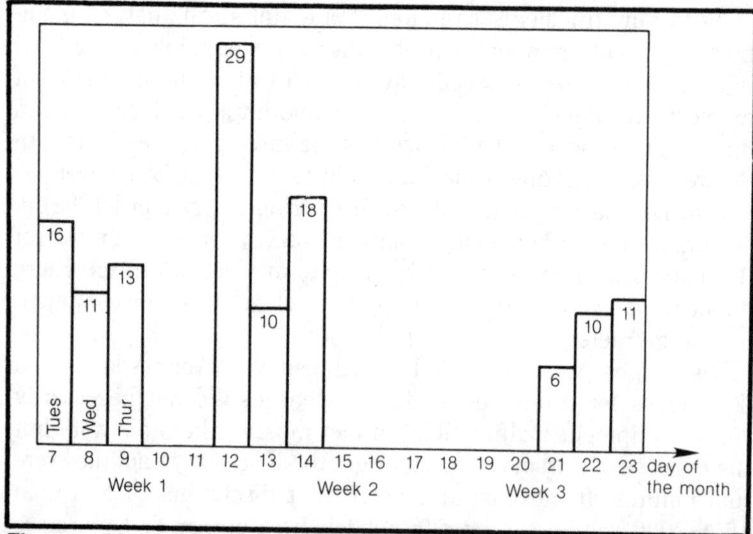

Figure 1.10—Number of discharges per day in an office area with one copier, one computer, one word processor, one telex, and one postage meter

- The purpose of the survey was to determine the number of *discharges, not the number of machine failures,* which would have restricted the scope of the study.
- To be sure that what was measured was the ESD current which people could generate (and not furniture ESD), and to provide every person with the same calibrated discharge path, a special monitoring set-up was devised:
  - I_{ESD} was measured by a current probe (current transformer) placed in a specially equipped metal doorknob.
  - Only the interior side of the door handle was equipped, so only personnel ESD from inside the room was recorded.
  - The current probe was connected to a recorder with 30 MHz bandwidth and 9 independent channels with specific threshold levels.
  - Each channel recorded the number of times its own threshold was *exceeded* (therefore can be directly translated into $P(I \geqslant I_x)$ statistics).
- The RH was constantly monitored.
- The campaign collected data from 498 eight-hour shifts, with about 120 discharges per shift. Therefore, about 60,000 ESD events were logged and arranged in 3800 data entries.
- According to sound statistical practices, a regression analysis weighted to the number of human contacts was performed.
- Over a more limited period, one type of anti-static carpet was also surveyed.

The 30 MHz bandwidth seems insufficient to measure ESD pulses with 1 nsec rise time (which would need a 300 MHz bandwidth). But the system was calibrated with a reference ESD pulser delivering a waveform with a 2 ns rise time, a 320 ns time constant, and a source impedance of 2000 Ω (behaving like a current source). Thus, the correlation between the 30 MHz limited bandwidth and the actual pulse bandwidth was taken care of by the calibration.

The analysis concludes, of course, that there is a strong correlation between the RH and the peak currents reached. A convenient equation for predicting the current, I, given a probability, P, was derived from the analysis:

$$I \geqslant 10^{A} \times P_o^{\,B} \times (RH)^{C} \tag{1.5}$$

where A=4.12, B=−0.645, C=−3.39 within the following range: 0.95 confidence, RH% comprised between 15-55, probability $P(I)=0.001<P_o<1$.

For example: What is the current which will be exceeded 10% of the time, given a RH% of 20, with 95% confidence?

$$I \geqslant 10^{4.12} \times (0.1)^{-0.645} \times (20)^{-3.39}$$

$$I \geqslant 2.2 \text{ Amp}$$

Eq. (1.5) has been arranged in tabular form in Table 1.2, while the results of the survey are shown in Fig. 1.11.

Table 1.2—Peak current (in amperes) having a probability P(I) of being exceeded

<RH≤ Percent		P(I) 1	0.5	0.1	0.01
15<	<20	0.22	0.80	2.4	8.7
20<	<25	0.14	0.73	1.9	7.1
25<	<30	0.11	0.35	1.1	3.5
30<	<35	0.072	0.20	0.80	2.2
35<	<40	0.059	0.11	0.40	1.3
40<	<45	0.052	0.075	0.17	0.56
45<	<50	0.042	0.055	0.10	0.25
50<	<55			0.072	0.15

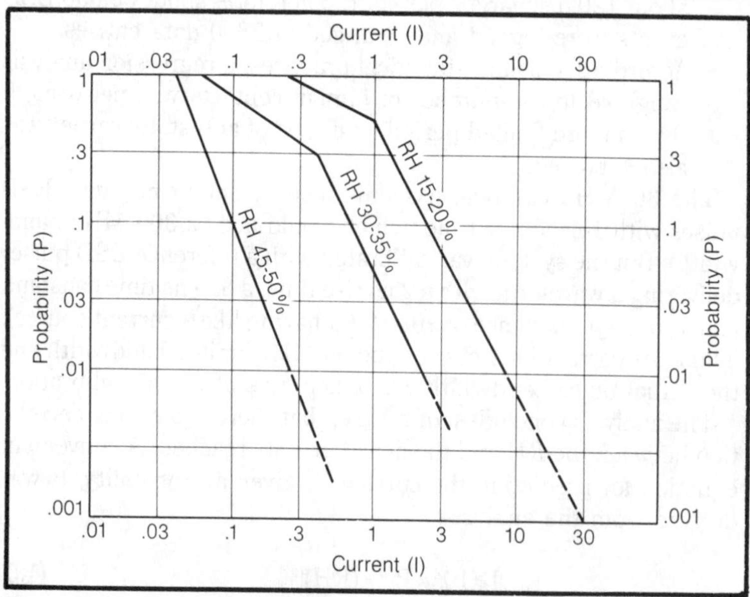

Figure 1.11—Probability (P) that a current (I) will be exceeded during personnel discharge

Table 1.2 shows that the current increases approximately like the inverse cube of RH. Given a same probability (P), currents will be 20 to 40 times greater at RH=15% than at RH=45%. Also, a given current I will be exceeded 100 times more often at a RH of 15% than at a RH of 45%.

The personnel ESD event curves give the current I which has a probability P(0.01<P<1) of being exceeded.

Since the RH has been recorded at intervals of 15-20, 20-25, etc., the curves shown are *mean values of P* in these intervals. Given the strong dependence of P(I) on relative humidity, even a 5% RH interval corresponds to large variations around the mean value of P for a given I. For instance, for an RH interval of 15-20, assuming the average RH is 17.5%, a current of 5 Amps will have a .025 (or 2.5%) probability of being exceeded. However, at the lower bound of this interval (15%), the current for the same value of (P) will be 8.4 amps, while at the upper bound (20%), the current having the same (P) will be only 3.2 amps.

To complicate the issue, the average value of RH in the interval *does not* correspond to the mean value of P(I) in that interval. Therefore, the curves of Fig. 1.11 are a good indication considering that, in reality, RH has daily fluctuations which often exceed 5%, but if a more accurate prediction is needed, Eq. (1.5) should be used.

Figure 1.12 shows a reconstituted histogram of the personnel ESD. On its lower scale, in addition to the current intervals, two voltages are shown:

- the voltage at which a 1 kΩ personnel simulator should be charged to replicate the same event
- the voltage at which an IEC-type simulator should be charged to create the same I_{ESD}

Not shown, but reported in Simonic's study, is the fact that antistatic carpet (acrylic carpet incorporating Brunslon® fibers by Brunswick) did show a 36-times reduction in ESD currents for the 15-20% RH interval and a 23-times reduction for the 30-35% interval.

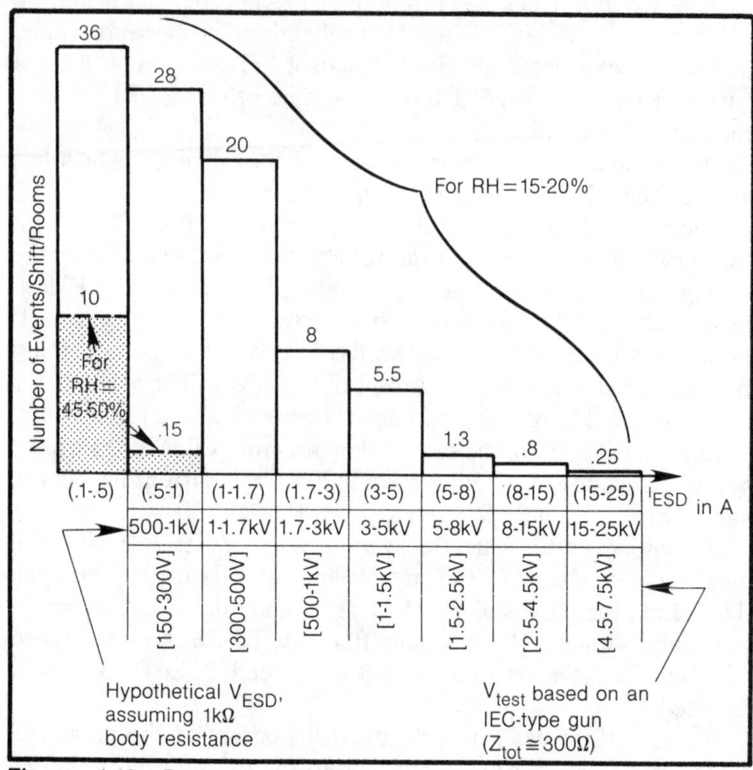

Figure 1.12—Reconstituted histogram—Events/shift for personnel discharge

1.4.2 Furniture and Objects ESD Statistics

The second of Simonic's studies covered furniture ESD events. The survey lasted several years and addressed ESD voltages in both computer room and data processing offices. The statistics are not compiled in probability but in number of ESD events per shift.

The highlights of the analysis are:

- The sites selected were locations with ESD problems, but where no corrective measures had yet been taken.
- The two kinds of sites were:
    - Computer rooms with raised metal floor and "humidity control": 10 sites, 11 machines, 3360 eight-hour shifts

- Carpeted offices with no humidity control: 8 sites, 8 terminals, 282 shifts
- All machines were floor-standing units with metal covers.
- The ESD event detector was a current probe placed around the I/O and power cables. The detector measured the peak current and was calibrated to correlate with a given discharge at typical contact points on the machine. The detector was also optimized for furniture ESD; i.e., voltage sources (low impedance metal objects). Therefore, the read-out was converted into an assumed ESD source voltage.*
- The detector 3 dB bandwidth was >100 MHz. As in the personnel study, the recorder had 9 channels with pre-set levels which counted each occurrence when the threshold was exceeded.

The philosophy behind this detector was that charged furniture behaves much like a voltage source; therefore, by knowing the average values of the discharge current path impedance, which includes the total loop resistance, the arc resistance and the loop impedance, the unknown ESD voltage can be derived as $V_{ESD} = I_{peak} \times Z_{loop}$.

A calibration set-up simulating the minimum (worst-case) loop impedance of the ESD discharge is shown in Fig. 1.13. Its dynamic impedance is comprised between 35 and 45 ohms; therefore, V_{ESD} (unknown)$=I_{peak}$ (measured)$\times$40 ohms.

Since the circuit is an RLC network with $R<2\sqrt{L/C}$, the discharge is an underdamped oscillatory waveform with a 10% to 90% rise time T, approximately equal to the charging time constant.

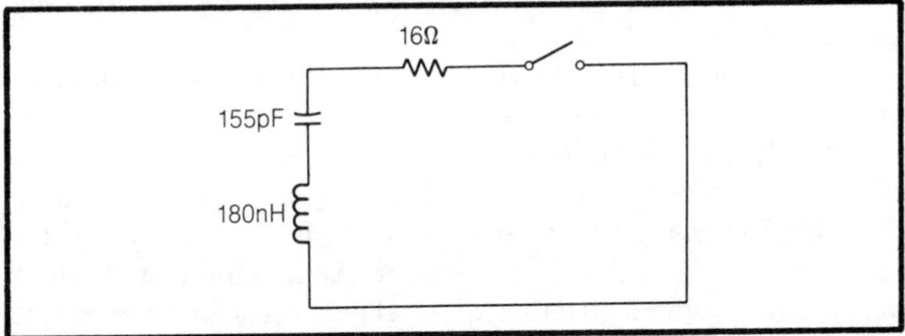

Figure 1.13—Reference discharge circuit used by Simonic to calibrate his furniture ESD monitoring (Ref. 11)

*The only questionable point would be the dependability of the conversion factor when the actual ESD waveform did not conform to the standard "template."

It is important to stress once more that this calibration assumes that the furniture discharge behaves like a voltage source with a discharge current dependent only on the load; i.e., the above discussed loop impedance. As a consequence, a personnel type discharge, with its true current, will be seen by the recorder, but the derivation of the actual ESD voltage would be wrong since the human body has a relatively high internal resistance. In contrast to the personnel discharge measurement, where the probe on the door handle prevented other-than-personnel ESD from being sensed, here the furniture ESD survey had no means of differentiating between actual furniture events and possible human events. A more thorough examination of the statistic curves will provide us, nevertheless, with a basis for this differentiation.

Figure 1.14 shows the event rate in the ten computer rooms surveyed. The curve labeled "mean" is the mean of all recorded events (all machines and seasons combined) weighted by the number of shifts monitored at each site. For instance, the likelihood of having a 1 kV furniture discharge (corresponding to about a 25 Amp discharge current) on a computer frame is about .13 per shift; i.e., about 50 times a year for 250 working days and 2 shifts/day. This does not seem catastrophic unless the dependability of the system is such that one error per week is intolerable. But one must remember that this is a *mean* figure. It is very likely that the majority of these events will be concentrated in the low RH period—January to March—which may mean there is actually one error per day! The threshold between what a computer user feels is tolerable and what is not is always subtle and is by no means a step function, but a system experiencing one error per day is generally unacceptable.

The table in Fig. 1.14 shows the spread of RH for both the computer rooms and the terminal rooms.

Figure 1.14 has a confidence interval of 95%. Using the previous example of a 1 kV furniture ESD, there is a 95% chance that more than .0025 events/shift exceeds that level, but only a 5% chance that it occurs more than .5 times/shift. Although the study does not specifically give a quantitative correlation between a given event rate and a given RH, it can almost be deduced from the data: There is a 25:1 ratio between the highest and lowest extremes. There is a strong chance that the highest voltages (>5 kV) were recorded

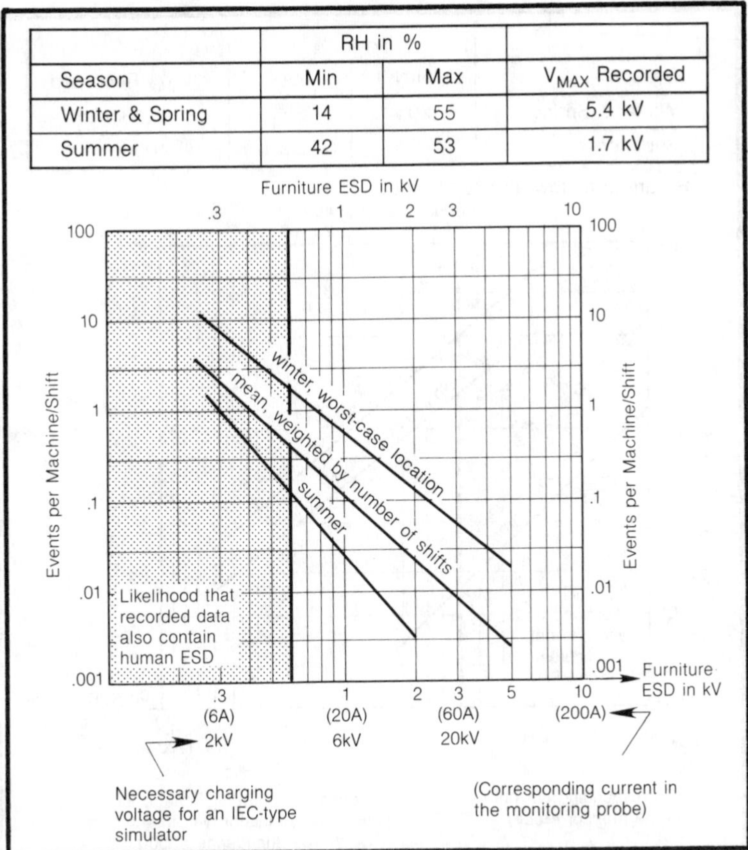

Season	RH in % Min	Max	V_{MAX} Recorded
Winter & Spring	14	55	5.4 kV
Summer	42	53	1.7 kV

Figure 1.14—Computer room furniture ESD events. Graph combines site dependency and season dependency (3360 shifts, 10 sites, 11 machines).

during the lowest RH periods, and the lowest during the highest RH period (although some low readings could occur during the dry season as well if a piece of furniture did not have time to yield a high electrostatic voltage before a contact occurred).

Figure 1.15 is even more revealing. The eight carpeted offices with work stations show a spread of E_{max}/E_{min}=5.6 kV/.09 kV=62 times. In the personnel statistics for carpeted offices, we had an equation showing that for a given probability P, the ESD current was varying like $(RH)^{-3.39}$ for $15 \leqslant RH \leqslant 55$. If we apply this relation-

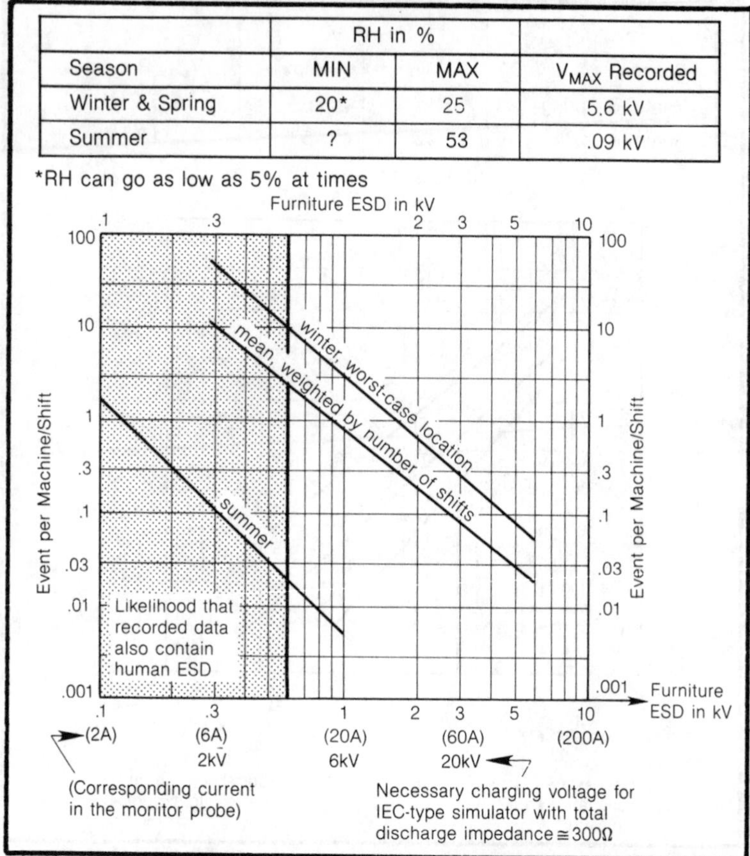

	RH in %		
Season	MIN	MAX	V_{MAX} Recorded
Winter & Spring	20*	25	5.6 kV
Summer	?	53	.09 kV

*RH can go as low as 5% at times

Figure 1.15—Terminal room furniture ESD events. Graph combines site dependency and season dependency (282 shifts, 8 sites, 8 machines).

ship for furniture ESD to the carpeted offices (and there is no reason to believe that carpets would not behave the same way), we find the predicted ratio of E_{max}/E_{min} to be 71 discharges, which is close to what it was in reality. The mean event rate for a 1 kV furniture discharge in these carpeted offices without RH control is about 1 per machine per shift.

When attempting to determine what portion of these ESD readouts may possibly be due to personnel discharges, keep in mind that the highest voltage a person can realistically build up and keep more than a few seconds is about 20 kV; with the worst-case human

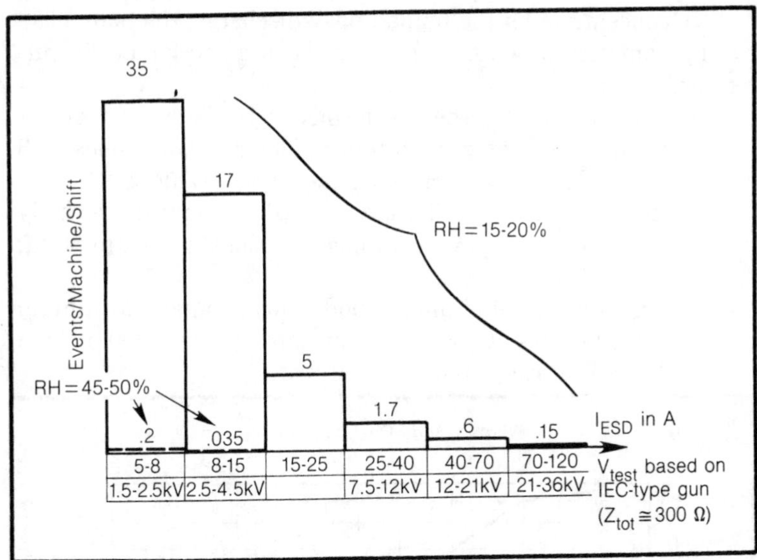

Figure 1.16—Reconstituted histogram—Events/machine/shift for furniture discharge in terminal rooms

body resistance being 1 kilohm, the greatest likelihood of personnel ESD data being inadvertantly mixed in with the furniture data would be I_p=20 kV/1 kΩ=20 amps, which would appear as 800 V on the event rate in Figs. 1.14 and 1.15. This has been shown as a shaded area on the left in both figures.

1.5 ESD Waveforms

1.5.1 Personnel ESD Waveform

In trying to match the measured waveforms and the idealized model of Sec. 1.2, simple waveforms for the ESD current have been devised with the following characteristics:

I = peak value of the current

τ_r = rise time of the current pulse, measured between the 10% and 90% points (approximately corresponding to the rise time of the triangle envelope)

τ = pulse width at 50% amplitude

If we concentrate on the human body discharge, the main electrical parameters which play a role in the rise and fall of the current are:

L = self-inductance of the loop formed by the body, its arm, the machine and the ground return. The range of values is .3 to 1.5 μH, with .7 μH being the typical value.

R_D = resistance of the discharge loop, dominated by body resistance. The practical range of values is 1 kΩ to 30 kΩ (Fig. 1.17).

C = capacitance of human body to ground. Minimum capacitance is 50 pF, maximum capacitance is 300 pF, and the typical capacitance is 150 pF.

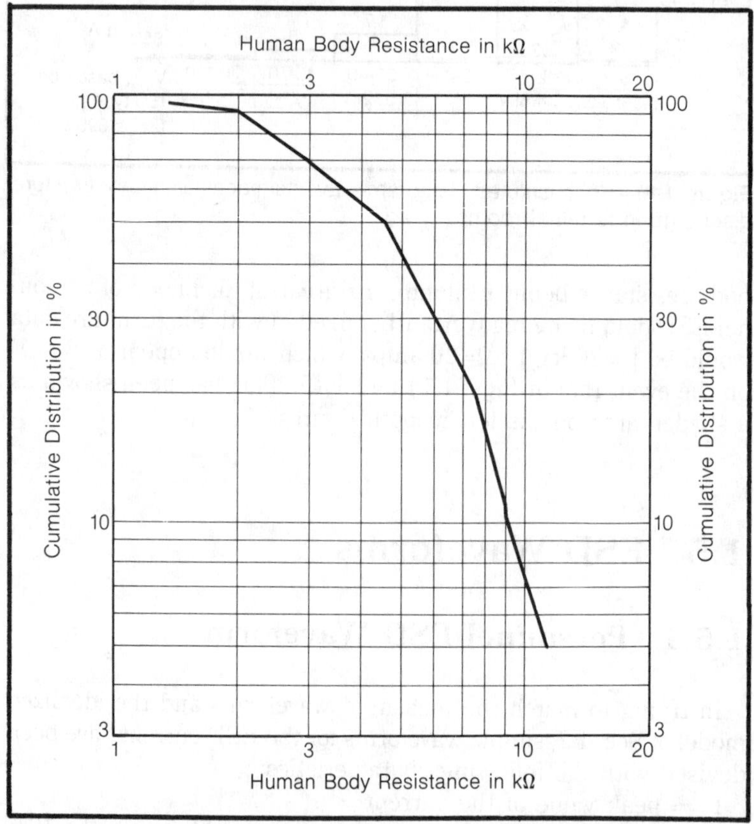

Figure 1.17—Distribution of human body resistances: probability (P) that R will be exceeded

The rise time of the discharge would be infinitely small if the capacitor simply discharged in a resistive network. In reality, the rise time is dictated by the charging time constant L/R_D. The pulse width of the discharge depends on the RC time constant of the circuit.

Table 1.3 gives a recap of the approximate range of rise times (τ_r) and pulse widths (τ), given all combinations of the extreme values for R, L, and C. They correlate rather well with actual measured waveforms. One could be tempted to take some average value to come up with a "standard" waveform. However, this is quite risky—an average figure for a normal distribution represents the

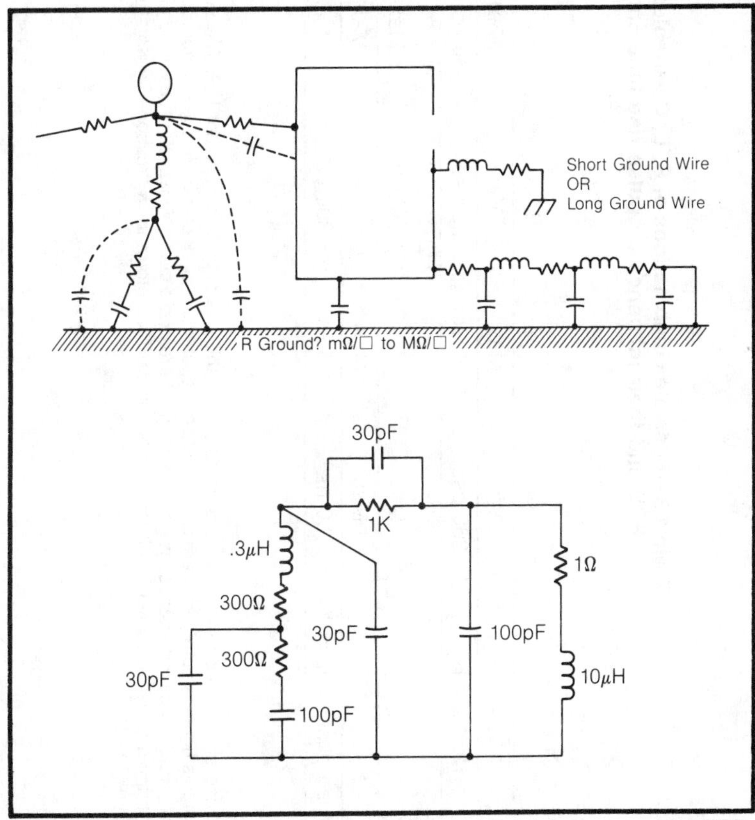

Figure 1.18—More complex lumped element model of personnel ESD showing how complex the phenomenon can be

Table 1.3—Possible combinations of R, L, C variables for personnel ESD and their influence on pulse rise time and fall time

Current Rise						
L	Min	Min	Max	Max	Typ	Typ‡
R_D	Min	Max	Min	Max	Typ	Min
Current rise $\tau_r \cong 1.2$ L/R*	360 psec	11 psec†	1.8 nsec	54 psec	240 psec	840 psec
Current Decay						
C	Min	Min	Max	Max	Typ	Typ
R_D	Min	Max	Min	Max	Typ	Min
Decay time constant $\tau_C = RC$	50 ns	1.5 μs	300 ns	10 μs	750 ns	150 ns

Notes:
*The 10-90% rise time is approximately equal to 1.2 times the charging time constant L/R. The 50% pulse width would be $\cong 0.69\tau_C$.
†The interest of this figure is purely academic. It would correspond to a peak current of a few hundred mA. However, dI/dt would still be there.
‡The spread of human body inductances is not very large; therefore, a standard waveform based on a typical value is justified. In contrast, the spread of human body resistances is huge, and a standard waveform for a "reasonable worst case" should aim to the lower bound of human body R_D.

value that is met in 50% of the cases and is *exceeded in 50% of the cases.* A specification designed upon this criterion would "under-protect" the equipment. It is safer to consider a reasonable maximum such as, for instance, the upper decile (the value which is exceeded in 10% of the cases only).

To come up with a reasonably severe waveform, let us look at the table showing the influences of R, L, and C and select the combination which provides the worst influence of each. For the human model, R_D is generally larger than L/dt, the inductive reactance of the loop. When L decreases the charging time constant decreases. When R_D decreases the peak current increases for a given static voltage, but this also slows down the rise time by changing the charging time constant.

Byrne (Ref. 13) has performed a thorough analytical study of human ESD by assimilating the body to a set of cylindrical shapes with their respective capacitances and inductances. He ends up with some low end extremes of 30 picoseconds for the rise time. Such short rise times have not been found in actual measurements. This does not mean they do not exist; displaying a 30 picosecond rise without distortion requires an instrument bandwidth of 10 GHz, which is not within the possibilities of memory oscilloscopes currently in use. However, such a discharge is associated with large values of R corresponding to smaller peak current values. Therefore, dI/dt, which is what counts for the radiation of the pulse to the victim, is fairly constant.

Fig 1.19 shows a waveform corresponding to a "standard" severe case, with a sharp 1 nsec rise time and a long exponential decay. A frequency spectrum of the pulse is also shown; since the pulse is a single event, its repetition period is infinite and there are no discrete spectral lines in the spectrum. The spectrum is a Fourier integral with its spectral density given in amperes per MHz of bandwidth. The spectrum starts (at a frequency equal to $1/\infty$) with an amplitude of $2 \times I_{peak} \times \tau$ μsec; the envelope is flat up to the first corner frequency F_1, then decreases like 1/F (20 dB/decade slope) up to the second corner frequency F_2, the reciprocal of the rise time which is often referred to as the "occupied" bandwidth. From then on, the amplitude rolls off like $1/F^2$ (40 dB/decade).

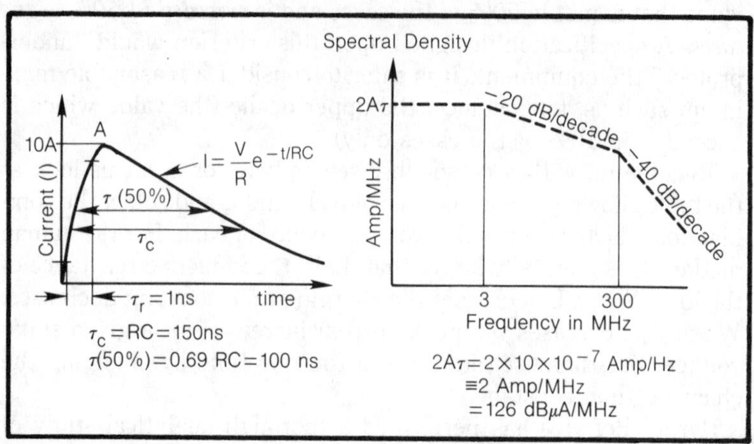

Figure 1.19—Personnel ESD current waveform and spectrum occupancy for a typical 10 kV discharge with R=1000Ω and C=150 pF

1.5.2 Furniture ESD Waveform

If we look now at the furniture or object discharge, the parameters are significantly different.

L = self-inductance of the loop formed by the furniture (a cart, for instance), the victim machine and the ground return. Values range from a minimum of .03 μH to a maximum of 1 μH. The typical value is .3 μH.

R_D = resistance of the discharge loop. This can be as low as a few ohms.

C = capacitance of the furniture or object to ground. This can vary widely from 30 pF to 500 pF.

Here the inductive part L/dt of the loop cannot be neglected versus R_D.

Now, compared to the human body discharge, the charging time constant has increased. On the other hand, the peak current will reach much higher values. Worse is that, instead of a slow discharging slope, we see a damped sine wave, typical of the ringing of an underdamped RLC circuit. Sec. 2.4 will discuss the impact of this ringing on the severity of the radiation coupling into nearby electronics.

Fig. 1.20 shows the waveform of a severe furniture discharge. Note that the ESD voltage at which the furniture was charged is less than that for the human discharge. This seems to contradict the assumption that furniture discharge is more severe. However, consider this: The furniture has a capacitance to the surroundings (i.e., the ground *and* the machine) which is typically larger than for the human case. This is simply due to the larger dimension of the conductive areas facing each other. A cart or metallic chair may have 3 or 5 times more capacitance than a human. Given that the quantity Q of electricity involved in furniture ESD is about the same as for human ESD (in fact, in most cases the furniture *has been charged from a human source* by charge transfer), the equation

$$Q=CV$$

shows that if C increases, the corresponding voltage has to be less for an equivalent energy storage.

Figure 1.20 also illustrates a frequency spectrum with the rise in spectral amplitude around the ringing frequency. Several well documented measurements, like those of King (Ref. 8) support this

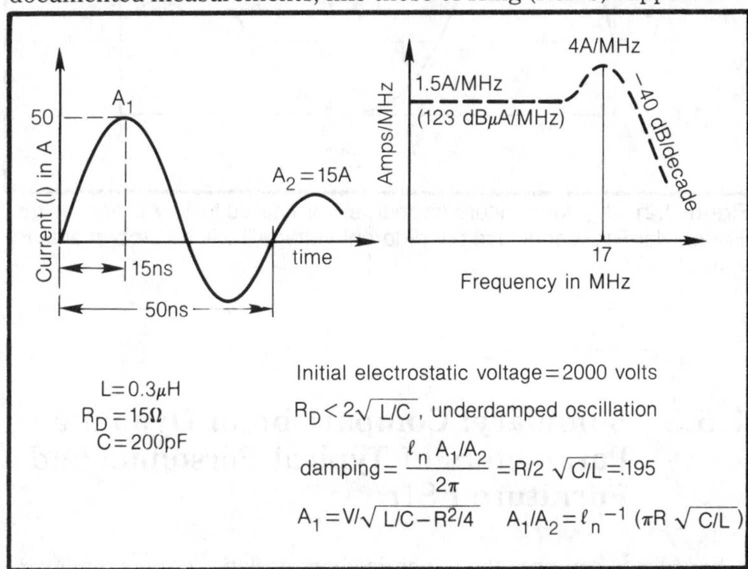

Figure 1.20—Furniture ESD current waveform and frequency spectrum for a 2 kV discharge

model of a low impedance ringing circuit. Together with Simonic's study (see Sec. 1.4) this tends to prove that the classical triangular-shaped pulse of human ESD is not enough to cover the variety of possible ESD events, and a furniture-type test with a discharge network having less than 50 Ω impedance would be a necessary supplement. To facilitate extrapolations, Fig. 1.21 shows a typical furniture ESD waveform referred to a 1 kV charging voltage.

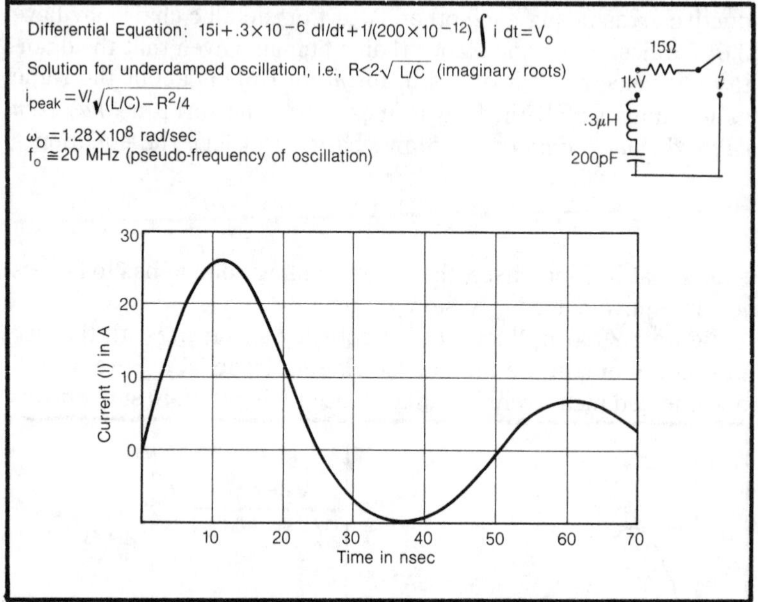

Figure 1.21—I_{ESD} for furniture discharge, normalized to 1 kV initial charge. (For similar RLC conditions but different voltage, I will be proportional to V_{kV}).

1.5.3 Summary: Comparison of Dynamic Parameters of Typical Personnel and Furniture ESD

In order to compare the dynamic characteristics of personnel versus furniture discharge, two typical waveforms (representing a severe, but not overly severe case) have been overlaid on Fig. 1.22.

An interesting result is that, given two ESD events having similar probabilities of occurring in a busy work space with uncontrolled RH, the personnel discharge is the one which seems to have the largest dI/dt and is therefore more likely to create the worst-case couplings. However, the duration of the steep change, i.e., the time during which the derivative exists, is 10 times larger for the furniture discharge than for personnel discharge. The important consequences of this will be discussed in Chapter 2.

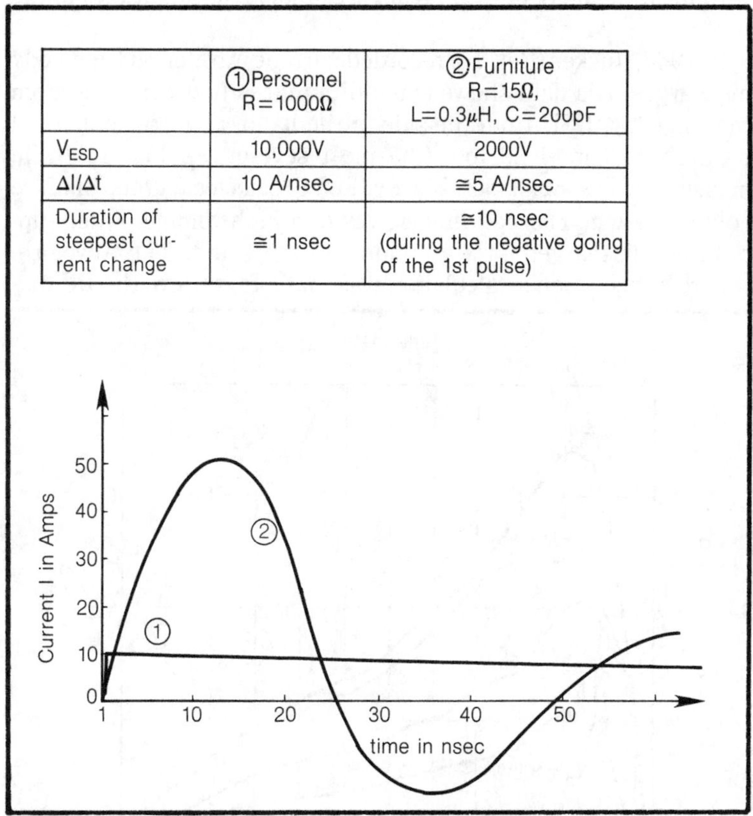

	①Personnel R=1000Ω	②Furniture R=15Ω, L=0.3μH, C=200pF
V_{ESD}	10,000V	2000V
ΔI/Δt	10 A/nsec	≅5 A/nsec
Duration of steepest current change	≅1 nsec	≅10 nsec (during the negative going of the 1st pulse)

Figure 1.22—Comparison of dynamic parameters of typical personnel **(1)** and furniture **(2)** ESD having the same probability of occurring in a severe environment (terminal room, winter season, P⩽1 event per shift). To allow comparison, probabilities have been related to discharge currents.

1.5.4 Actual Versus Idealized ESD Waveforms

No two electrostatic discharges look alike. Therefore, pretending that the personnel or furniture ESD waveforms shown in Figs. 1.19 and 1.20 are close to what actually happens would be presumptuous. The idealized waveforms have been devised as a repeatable test criterion; that is, an equipment which resists such standard ESD waveforms is likely to resist real ESD of similar amplitudes in the field.

In 1968, Tucker (Ref. 14) recorded current waveforms from body discharges. His data shows some differences between ESD from the fingertip, ESD from the side of the hand, and ESD enhanced by a sharp hand-held tool (the most severe; see Fig. 1.23). To measure ESD waveshapes, Mazdy (Ref. 9) selected a group of brave volunteers who charged themselves to a high-voltage power supply then discharged on a grounded 1 ohm shunt. The waveform was recorded, then a calculation was made to retrieve the RC net-

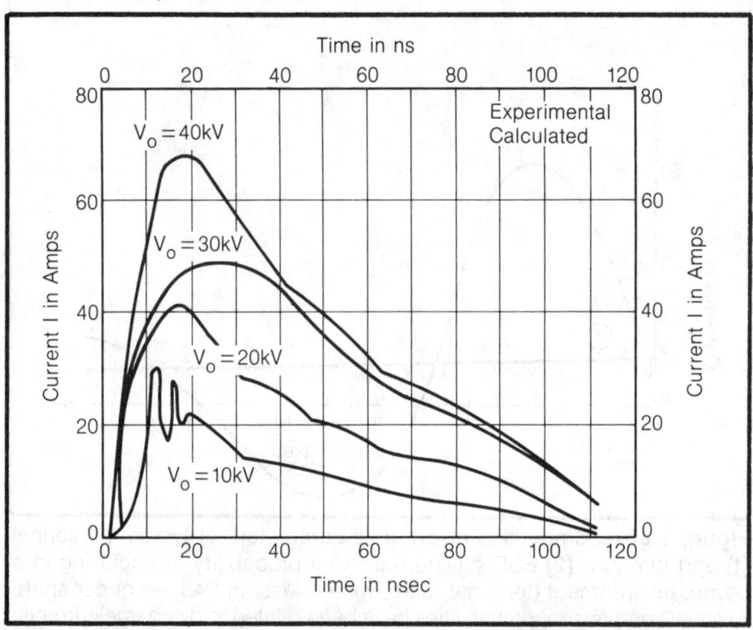

Figure 1.23a—Personnel ESD waveforms, recorded with one artificially charged person (see Tucker, Ref. 14). This graph shows ESD from a hand-held metal tool.

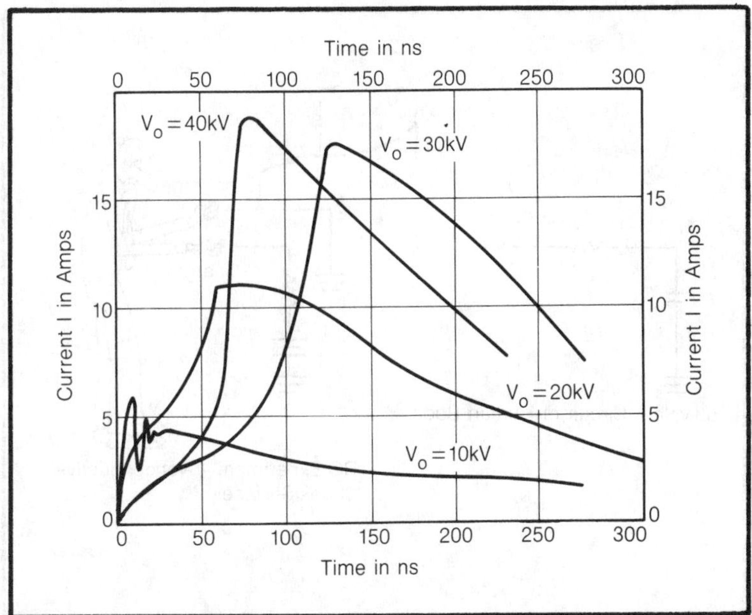

Figure 1.23b—Personnel ESD waveforms, recorded with one artificially charged person (see Tucker, Ref. 14). This graph shows ESD from a fingertip.

work which best fit the actual data. Fig. 1.24 shows a summary of the results. No inductance was put into the model because the study was mainly aimed at determining the destructive effect of ESD when handling modules; i.e., the pulse duration was the concern, not the rise time.

However, at a time when no standard existed for ESD immunity of integrated circuits, Mazdy's data and the simulator he built using them was widely used by IBM for the QC of its modules.

Around the same time, King (Refs. 7 and 8) began a thorough study of ESD waveshapes involving mainly personnel, except in one case where an oscilloscope cart pushed against the discharge set was used. The bulk of King's data for personnel is shown in Fig. 1.25.

One sees that with the classical personnel discharge, the finger approaching slowly as ionization occurs, the measured waveforms

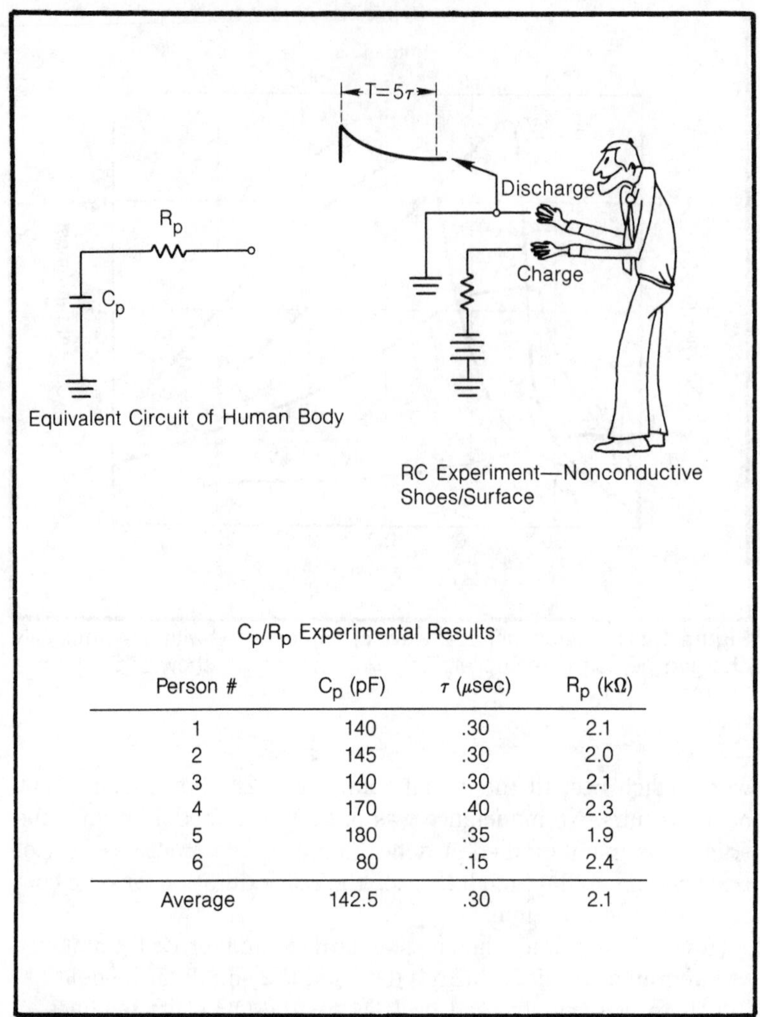

C_p/R_p Experimental Results

Person #	C_p (pF)	τ (μsec)	R_p (kΩ)
1	140	.30	2.1
2	145	.30	2.0
3	140	.30	2.1
4	170	.40	2.3
5	180	.35	1.9
6	80	.15	2.4
Average	142.5	.30	2.1

Figure 1.24—Experiment to determine equivalent RC circuit of human body and results (see T. Madzy, Ref. 9)

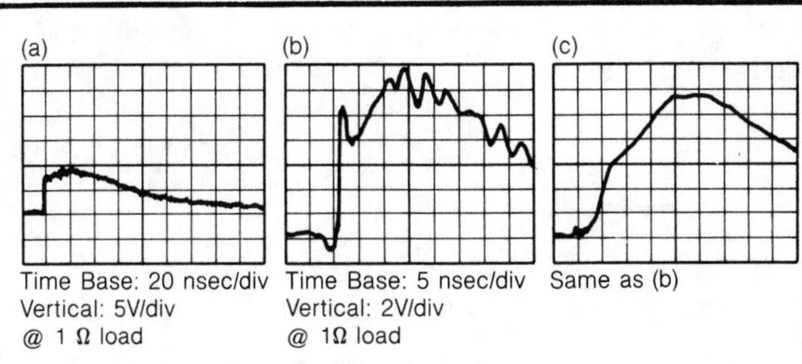

Measurement Conditions: The above data were taken under the conditions of

- Initiating level: 5,000 Volts
- Human subject holding metallic intervening object (screwdriver) as the ESD path
- Motion: Discharge load slowly approached to allow maximum ionization to occur.

V_{volts}	Extremes of I_{ESD} range (Amp)	I_{ESD} aver.	$Z_{aver.} = V/I_{aver.}$
500	.9 to 1.8	1.3	385Ω
1000	1 to 3.6	2.3	435Ω
2000	4 to 5.8	3.5	570Ω
4000	1.8 to 7.6	4.25	940Ω
6000	1.2 to 26	10	600Ω
8000	1.8 to 8	6	1250Ω
10,000	4 to 6.5	5.2	1920Ω

Figure 1.25—A few of the personnel ESD waveforms recorded by King (Refs. 7 and 8) using artificially charged "volunteers" and enhanced discharge (screwdriver). Figures (b) and (c), taken with expanded scales, show large variations (1 to 15 nsec) between rise times for different persons and attitudes. A compilation of I_{ESD} vs. V shows clearly an increase of Z with voltage. This non-linearity of the discharge circuit can be explained (to some extent) by the increasing impedance of the arc.

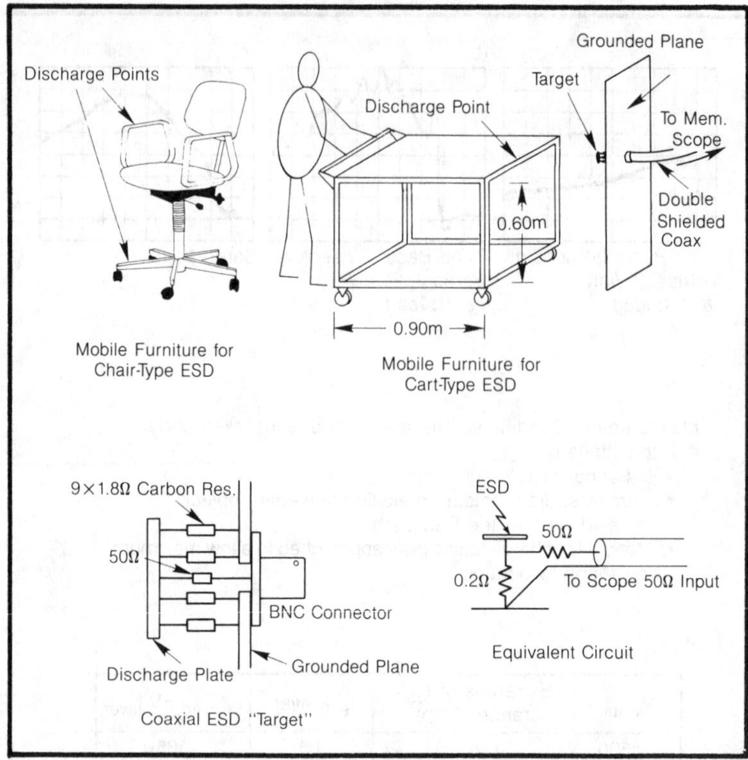

Figure 1.26—Test set-up used by King (Refs. 7 and 8) for his furniture discharge waveform study. For both the chair and the cart, the person was initially charged through a dc power supply.

are fairly close to the idealized waveshapes of Sec. 1.5.1. In contrast, when the field is locally enhanced by a sharp tool* or the finger is approached very rapidly, the rising edge shows some odd shapes, with a "precursor" current spike having only a few hundred picoseconds of rise time. This is somewhat reminiscent of lightning where, too, a precursor is often followed by one or several restrikes. King and other ESD pioneers attribtue this precursor to a localized generator where the hand (and eventually the hand-held tool) capacitance to grounded object is discharged first, and so fast that this segment of the body is "disconnected" from the rest. Then

*This would also occur with a wristband, rings, coins, keys, etc.

the path deionizes for a few hundred picoseconds until the rest of the charge left behind reaches the finger area and restarts the main discharge. At low levels, the surface-to-surface distribution between hand/finger and the "load" creates the precursor current spikes. At higher levels, the increased path inductance (a 1 cm arc has about 10 nH of inductance, presenting 100 ohms to a nsec current charge) slows down the rise time and permits a more complete transfer of the whole body charge, without the "contact bouncing" seen at lower levels. The transition between low and high level behavior occurs around 6-8 kV (a phenomenon also shown in the earlier measurements by Tucker, Ref. 14).

In a follow-up of these experiments, King and Reynolds (Ref. 8) addressed furniture discharge. Their main findings are shown in Fig. 1.27. The decaying oscillations typical of an RLC circuit with $R<2\sqrt{L/C}$ (underdamped) are clearly identifiable. The peak currents are impressive as well. King's experiments were conducted with many precautions concerning the high frequency response of his set-up; for instance, a copper ground plane was used to avoid

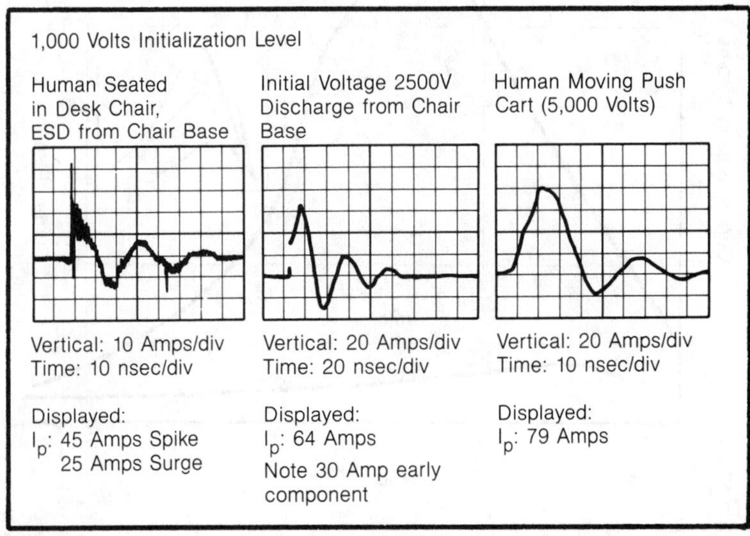

Figure 1.27—A few of the 50 furniture ESD waveforms recorded by King (Refs. 7 and 8). Large variations were experienced due to small changes in people's attitudes and speed of motion towards the target. The pre-discharge phenomenon is visible, as in personnel ESD, but it disappears above ≅2500 V instead of 6000 V for personnel.

uncontrolled parasitic inductance in the return path. The shunt used to read the current was a coaxial mount, and the instrument 3 dB bandwidth was larger than 500 MHz.

More recently, Hyatt (Ref. 12), Ryser (Ref. 15), and Richman (Ref. 16) closely studied this pre-discharge and the way to simulate it. They found that the dI/dt during the pre-discharge can be as high as 30 A/ns if a finger is approached fast enough, and the phenomenon can be modeled very exactly with an additional LC element replicating the human hand. These oddities introduce some non-proportionality between V_{ESD} and I_{Peak}. For instance, if a 15 kV initialization creates a 15 amp peak current, it is not certain that a 5 kV level will create a 5 amp current.

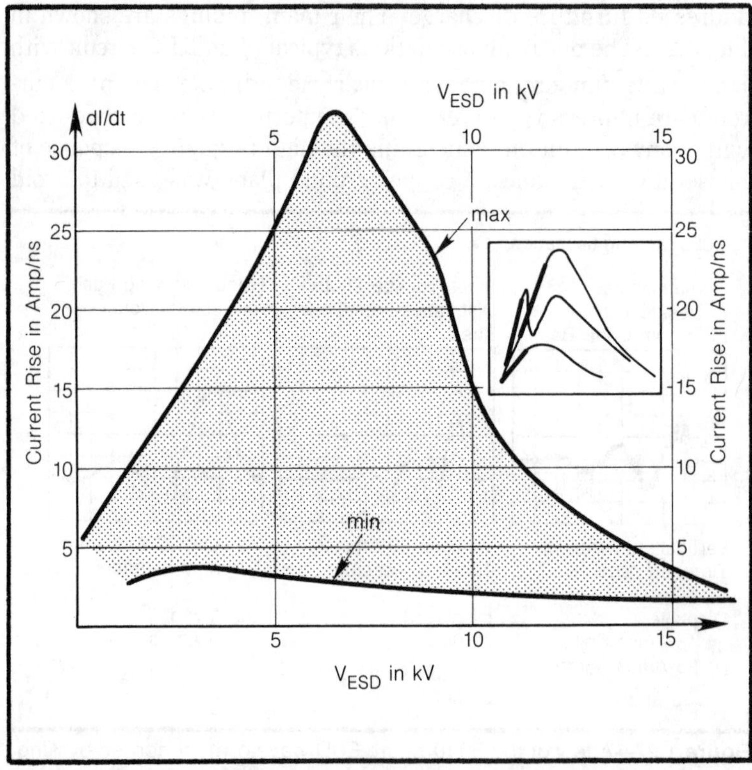

Figure 1.28—Range of dI/dt reported by Ryser (Ref. 15) depending on the speed of approach of an ESD probe. On top right corner, see the ambiguity in dI/dt caused by the "precursor" and other odd waveshapes.

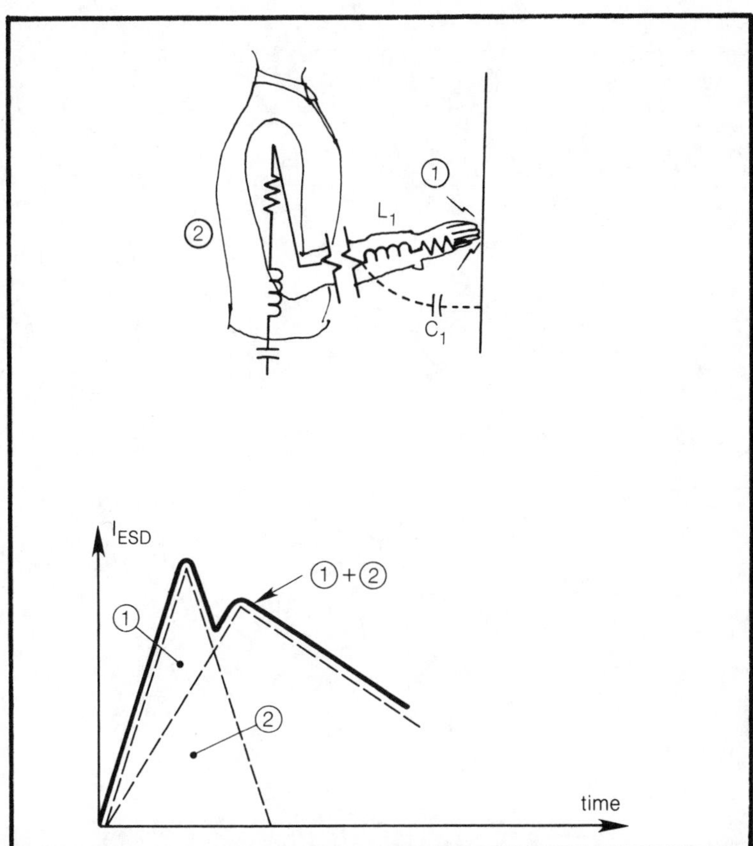

Figure 1.29—The proposed explanation for the precursor discharge (see Ref. 16). C_1 and L_1 being very small, the rise time and duration of **(1)** are very short.

Chapter 2

ESD Coupled into Electronics

Not accounting for the multitude of ESD incidents in the modern environment and concentrating only on ESD's impact on electronic devices, the ESD problem can be broken down into three essential scenarios:

- Direct discharge to an electronic component (integrated or discrete)
- Direct discharge to an electronic equipment housing
- Indirect discharge

2.1 Direct Discharge to an Electronic Component

Direct electrostatic discharge to an electronic component is one of the most severe threats to component reliability during the manufacturing and handling of electronic parts. It costs *millions* of dollars of losses in dead chips and severe breeches in the percentage yield of IC manufacturing lines, not to mention the latent wounds which will later show up as unexplained failures in the field.

This problem was recognized in the late 60's, and all the large manufacturers of electronic components have now implemented strict static control programs comprised of:

- Static awareness for employees
- Static-free work areas and manufacturing hardware
- Monitoring of percentage yields and chip infant mortality to detect ESD symptoms
- Anti-static precautions for card handling in the field.

This book does not address component failure but rather the malfunctions of a complete equipment. The reader who is interested in the component failure aspect of ESD is referred to the abundant literature existing on the subject (Refs. 1, 2, 5 and 6). We will simply give a brief reminder of this aspect of ESD by showing a few typical episodes.

In Fig. 2.1a, the module is on a workbench and is "zapped" by

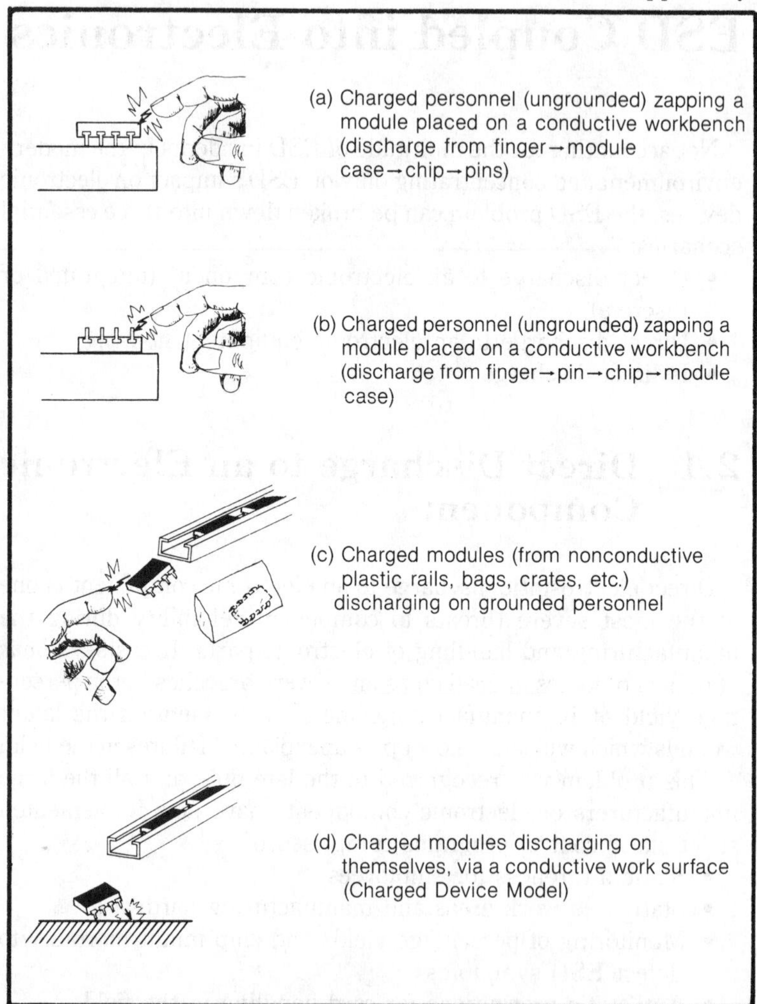

(a) Charged personnel (ungrounded) zapping a module placed on a conductive workbench (discharge from finger→module case→chip→pins)
(b) Charged personnel (ungrounded) zapping a module placed on a conductive workbench (discharge from finger→pin→chip→module case)
(c) Charged modules (from nonconductive plastic rails, bags, crates, etc.) discharging on grounded personnel
(d) Charged modules discharging on themselves, via a conductive work surface (Charged Device Model)

Figure 2.1—Integrated circuit damage during handling

a person who touches it. Note that:

- The culprit person may not even notice the ESD event if he or she was charged at less than 1.5-2 kV (threshold of feeling); therefore, no warning exists of possible damage.
- The sad thing is that if the workbench was conductive and grounded (a good anti-static practice), this would only make things worse by decreasing the impedance of the ground path.

The moral is that anti-static precautions should be applied across the board. Nobody should handle modules without using a grounded wrist-strap or, at the very least, touching a grounded structure first; the workbench should be conductive, yes, but "softly" grounded via a few hundred kilohms resistor.

Plastic encased modules are prime victims of this type of event since the field gradient creates arcing or charge transfer from the case surface to the chip itself, then sinks to ground via the leads. Metal encapsulated modules have better chances of survival if the arc flows on the metal can and breaks the can-to-benchtop air gap.

Figure 2.1b shows the opposite situation where the module is standing with the leads facing upwards. Here metallic-canned modules may be as badly damaged.

Figure 2.1c shows a more insidious scenario: The module has become charged by sliding in carousel rails or by being put in ordinary plastic bags, crates, skin packing, etc. The device can remain charged a significant amount of time until it finds an occasion to get rid of its excess (or lack) of electrons, hence receiving a self-inflicted wound.

Few reports exist of the actual current waveforms in this latter case, but given that the impedance of the device can be very low, and that the benchtop may be conductive (for the very purpose of avoiding static!), it is assumed that the peak current and the total energy can be significant. This scenario, known as the "Charged Device Model," has been identified and discussed in several papers (Refs. 17 and 18). The solution is to always ship ESD sensitive devices in anti-static bags, preferably the shielded type, and to have their pins shorted-out in a conductive foam pad.

Remember also that all the above scenarios repeat themselves throughout the whole life of a device:

- During wafer processing, test, and cutting
- During chip transport, test, substrate mounting and encapsulation

- During module handling, test, and storage
- At the next step; i.e., PCB level (contrary to common myth, an IC mounted on a PCB *is* still vulnerable)
- In the host machine itself, during maintenance, etc.

Fig. 2.2 shows a typical static free work station, and Fig. 2.3 shows some elements of static protection at the factory.

What ESD level does it take to damage a component which is at rest and unpowered? The answer, of course, depends on:

- The type of technology
- The type of built-in protection provided by the manufacturer (assuming that the clamping diodes are themselves ESD resistant, which is not always true)
- The dimensional rules: Silicon oxide thickness, width and thickness of traces, traces and pads spacing.

ESD failures generally occur by upsetting the breakdown voltage of the SiO_2 isolation. For instance, in MOS structures with typical oxide thicknesses of 1000 Å (10^{-7} meter), the rigidity of 5-7 MV/cm is exceeded if a voltage of 50-70 V is reached between two isolated channels. Another common failure is by the joule effect where a

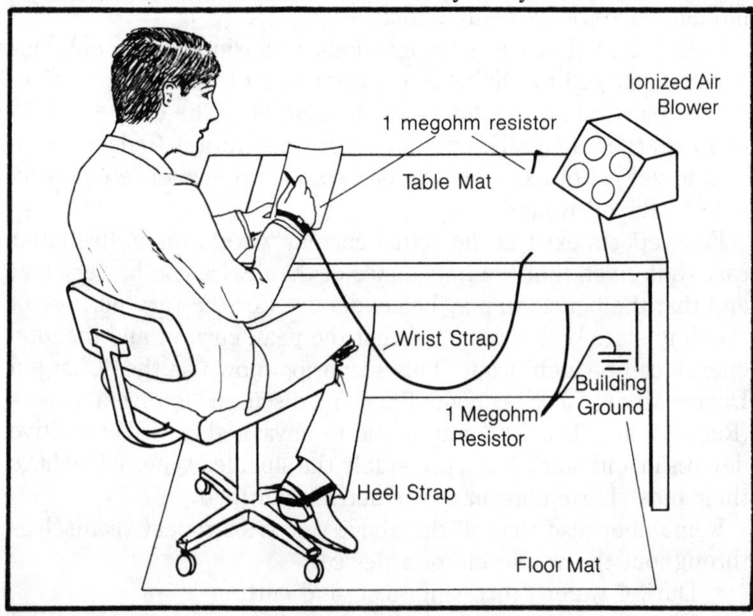

Figure 2.2—Example of a static-free work station

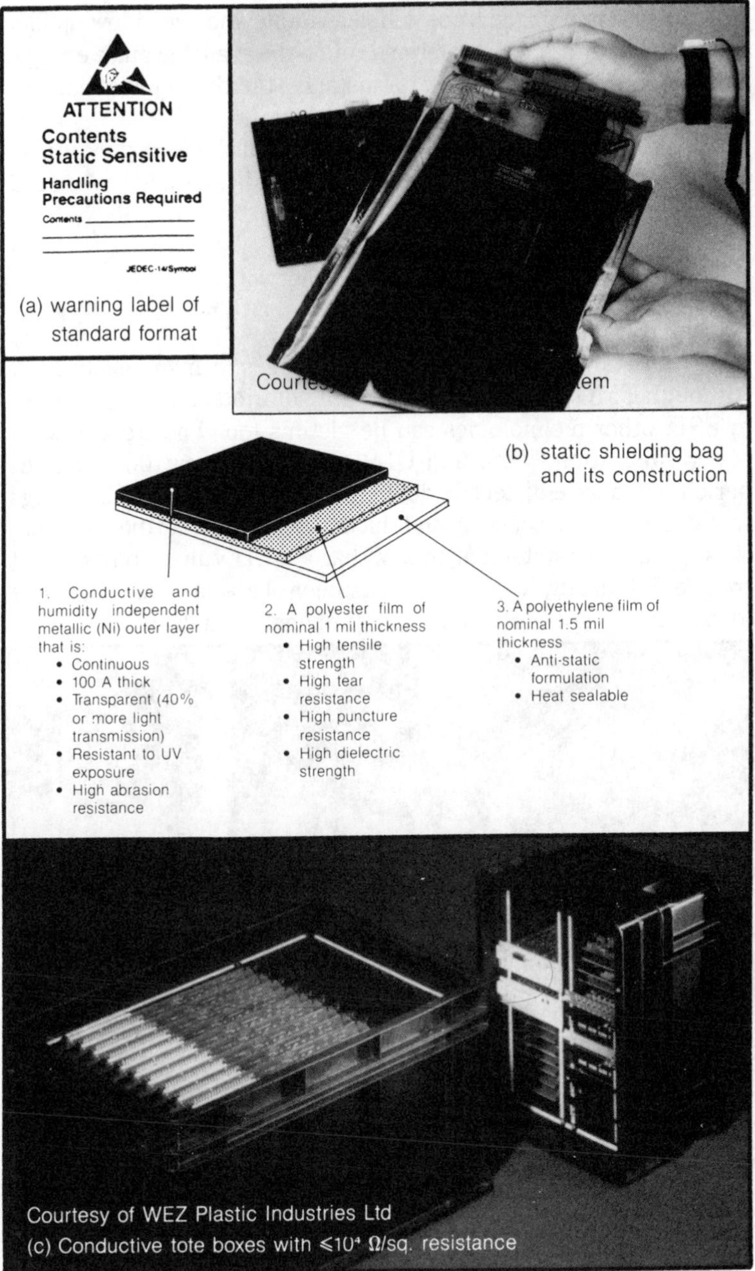
ATTENTION
Contents
Static Sensitive
Handling
Precautions Required
Contents
JEDEC-14/Symbol
1. Conductive and humidity independent metallic (Ni) outer layer that is:
• Continuous
• 100 A thick
• Transparent (40% or more light transmission)
• Resistant to UV exposure
• High abrasion resistance
2. A polyester film of nominal 1 mil thickness
• High tensile strength
• High tear resistance
• High puncture resistance
• High dielectric strength
3. A polyethylene film of nominal 1.5 mil thickness
• Anti-static formulation
• Heat sealable
Courtesy of WEZ Plastic Industries Ltd

(a) warning label of standard format

(b) static shielding bag and its construction

(c) Conductive tote boxes with ≤10^4 Ω/sq. resistance

Figure 2.3—Static protection at the factory

junction or a metallized trace will deteriorate and even blow up like a fuse. In either case, the criterion is often the available pulse energy. For current MOS technology, for instance, the threshold of damaging energy is on the order of 1 microjoule.

Since Energy=Power×Time, and given that the 50% pulse duration of human ESD is about 150 nsec, this indicates a not-to-exceed pulse power of:

$$P = \frac{10^{-6}\ J}{150 \times 10^{-9}\ sec} = 6.6\ watts$$

Fig. 2.4 shows some typical ESD damages in microelectronics. Although CMOS and FET devices are notorious for their fragility to ESD, other technologies can be victims too. The trend toward hyper-integration (VLSI and ULSI with 10,000 or 100,000 gates per chip and gate dielectric down to 250 Å) will only make things worse since all the spacings and thicknesses will be further reduced. It is not an overstatement to say that if ESD vulnerability is not treated drastically, it can become a major obstacle in the challenge toward faster speeds and shorter propagation delays.

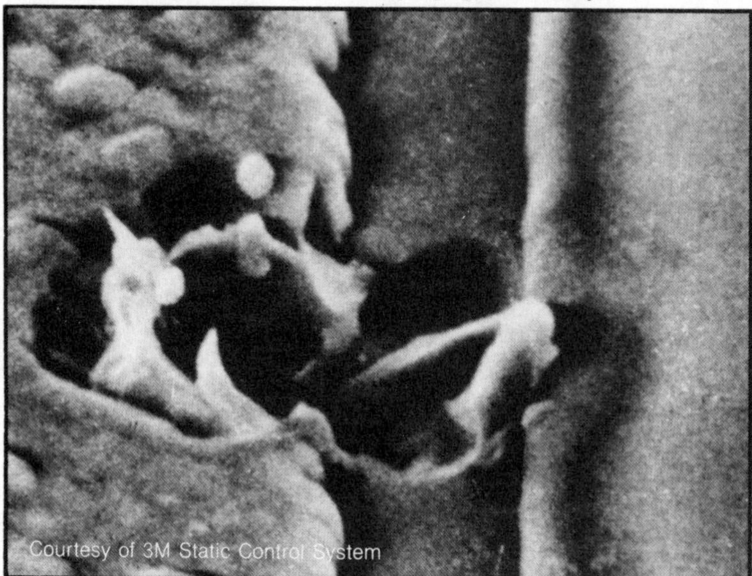

Figure 2.4—Magnified view of integrated circuit damaged by ESD

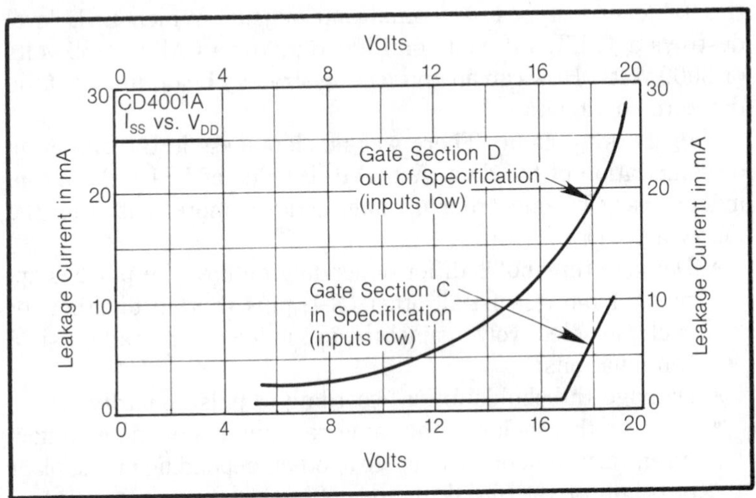

Figure 2.5—Example of deterioration in reverse breakdown (V, I) characteristics. V_{BR} is often used as an indicator of ESD overstress, either by monitoring V for a given current, or monitoring $I_{leakage}$ for a given voltage. The parameter surveyed here was the quiescent leakage current I_{ss} on a CMOS NOR gate. (From Ref. 6b.)

Table 2.1—ESD susceptibility of various electronic devices

Device Type	Range of ESD Susceptibility (Volts)
VMOS	30 to 1800
MOSFET	100 to 200
GaAsFET	100 to 300
EPROM	100
JFET	140 to 7000
SAW	150 to 500
OP AMP	190 to 2500
CMOS	250 to 3000
Schottky Diodes	300 to 2500
Film Resistors (Thick, Thin)	300 to 3000
Bipolar Transistors	380 to 7000
ECL (PC Board Level)	500 to 1500
SCR	680 to 1000
Schottky TTL	1000 to 2500

Courtesy of 3M Static Control System

Table 2.1, often reported in the literature (Ref. 6), shows the susceptibility of various electronic devices to a standard human body ESD.

For someone used to dealing with a certain degree of accuracy, this table may appear as some kind of joke. Which ESD level destroys a JFET: 140 volts or 7000 volts? or a CMOS: 250 volts or 3000 volts? How can any protection strategy be optimized if the data are so vague?

Nobody is to blame. There was no sloppiness in the gathering or compilation of the data. Actual thresholds of ESD failure can, indeed, vary by one order of magnitude or more, based on the following:

- Damage thresholds differ depending on how the pulse is applied: 1 pin against all other (N-1) pins shorted together? or each pin selectively against the grounded case? or all possible combinations?
- Damage thresholds differ according to pulse polarity
- Damage thresholds of the same part number, same manufacturer, vary from one vintage to another, depending on the plant of origin or the serial number of the masks used

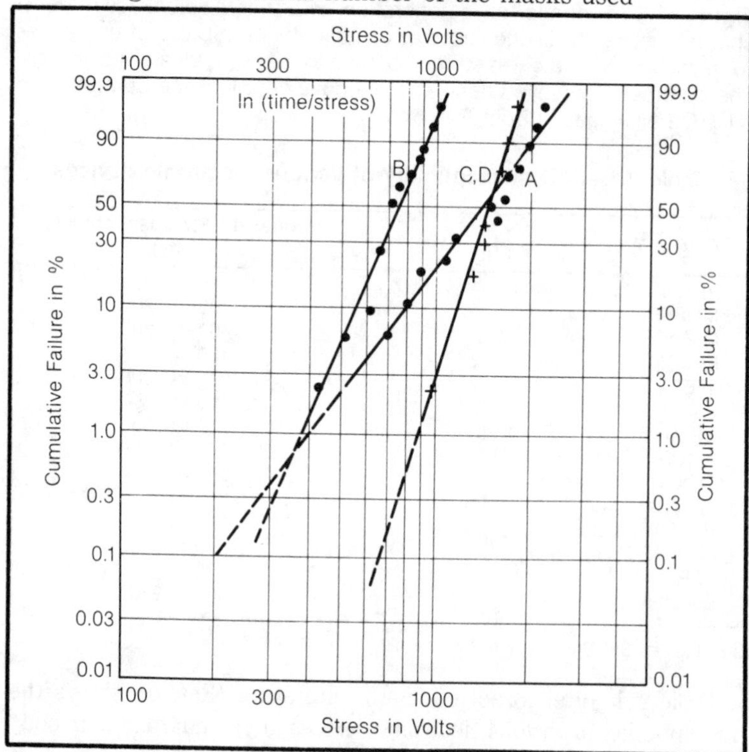

Figure 2.6—ESD failure levels for 4001 inverters from manufacturers A, B, C and D

- Damage thresholds are not the same for a unique test pulse as for repeated pulses
- Damage thresholds of identical, compatible part numbers from different manufacturers vary widely
- Reported levels depend on the criteria selected: Was a part declared failed when it was functionally wrong? or as soon as its dc parameters deviated from the initial specification?

Fig. 2.6 from Ref. 22 shows, on a Weibull distribution graph, the ESD threshold of CMOS inverters from four different manufacturers. The monitored parameters included an input current >1 μA (indicating a damaged gate isolation) or output not responding to input change, whichever failure came first.

The graph shows ratios of 3 to 1 or 4 to 1 between the best part and the worst part of the sample. Also, the deviation varies widely from one manufacturer to another.

Another example showing that the ESD damage level is never a "green light/red light" situation is shown in Fig. 2.7 from Ref. 19. The product tested was a bipolar logic gate type 54 L04 (TTL Low Power). One hundred devices were tested for multiple pulses until a failure occurred.

The criterion for failure was the deviation of two critical parameters from their normal value. The normal value of $I_{input\ High}$ was 10 μA_{max}, and failure was declared if Δ>20 nA. The normal value of $V_{out\ Low}$ was 0.3 Volts max; failure was declared if Δ>.05 volts. The results are shown in the number of parts that did not survive to N pulses at level V.

Pulse Level	Number of Pulses to Failure									
	1	2	3	4	8	50	75	125	175	200
5750	10									
5000	12	1			2					
4250	16				1					3
3500	9			2	1					8
2750	2		1		1	1	1	1	2	11
2000										10

Figure 2.7—Failure decision is never a "step function"! Example: damage testing of a 54 L04 (TTL Low Power). One hundred devices were tested. Normal specification of $I_{input\ high}$ was 10μA max, and the device failed if Δ>20 nA. Normal specification of $V_{out\ Low}$ was 0.3 volts max, and failure was declared if Δ>.05 volts.

All the parts survived after the application of up to 175 pulses at 2000 V, then ten died between 176 and 200 pulses. The survivors were stressed to 2750 V, where casualties are seen as soon as the first pulse. Finally, of the 15 survivors which came victoriously through the 200 pulses at 5000 volts, 10 died at the first application of a 5750-volt electrostatic discharge. Can we then say that we have a product that can withstand ESD at 2000 volts? or up to 5000 volts?

This is a case where all the usual methods of stress tests and QC statistics have to be used to determine a reliable number which will characterize the ESD immunity with a certain confidence level. In all these tests, an unstressed sample lot should be monitored to avoid introduction of an uncontrolled variable not related to ESD.

An interesting question arises: Would not the 15 heroes that came through 5000 V have gone even further if they had not been inflicted at lower levels 600 discharges before? Sometimes, to investigate this, fresh parts, marked for identification, are reintroduced into the lot to replace the victims as the test goes on.

Another related problem is one of latent failures. A part which still *appears* undamaged after an ESD test may, in fact, have its lifetime affected. There are several confirmations that parts which have been subjected to ESD (either by accidental or intentional events) become "walking-wounded" (see Pete Richman, Ref. 4) and exhibit abnormal failure rates in the field. But even this is not always true—some sample lots which were ESD stressed have exhibited, during accelerated life tests, better lifetimes than unstressed sample lots! This self-healing phenomenon has been given several explanations, the descriptions of which would be far beyond the scope of this book.

Since 1980, the U.S. Department of Defense has issued a standard document, DOD-STD 1686, defining all the requirements of an ESD control program for electronic components and assemblies. This standard, originally intended for suppliers and subcontractors of the DOD, has become a commonly used reference for industry in general.

The DOD standard calls for:

- identification and tagging of ESD-sensitive items (class 1 with sensitivity ⩽1000 V and class 2 with sensitivity between 1000 and 4000 V). Details of the test are given in Chapter 3.
- built-in circuit protection at chip and card level
- ESD-proof handling, shipping, etc. procedures
- QC and audits
- field maintenance precautions

The standard also requires that subcontractors rule out class 1 devices when a class 2 device is available which would perform identical functions.

As a complement to DOD-STD-1686, the Department of Defense has also issued Handbook #263 which gives ESD control guidelines and details of failure mechanisms. The handbook also contains classifications of ESD protection equipment, materials, manufacturing, and shipping procedures. (For an excerpt from MIL-HDBK-263, see Appendix A.)

2.2 Direct Discharge to Electronic Equipment Housing

The direct discharge is the most classical case and is the easiest to understand. The charged person or object (the "source") touches a metal enclosure, the "load." Most of the time (Fig. 2.8a), the discharge occurs on a purely mechanical part which is touched in-

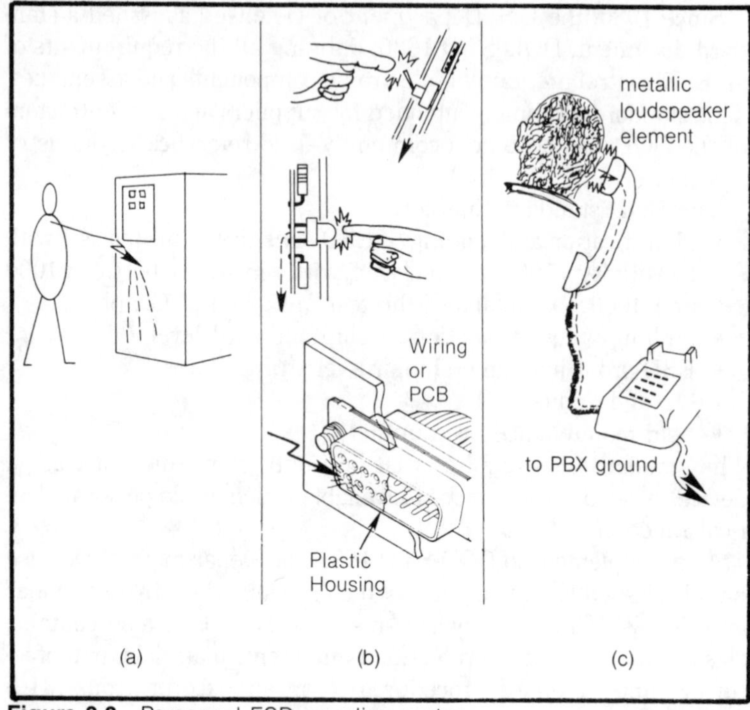

Figure 2.8—Personnel ESD coupling routes

tentionally (knob, key, switch, handle) or accidentally (frame, covers).

More severe occurrences, like those shown in Fig.2.8b, include:

- finger approaching an unprotected I/O connector
- finger arcing through an LED or incandescent display
- discharge on a PCB mounted switch, in which case a subsequent arc occurs *internally* between the toggle and the active contacts of the switch.

In these cases, the ESD current can reach the electronic components directly via a conducted path. Except for some damping caused by the wire or trace length, the situation is almost as severe as the direct discharge to a module pin. This is discussed in Sec. 2.1.

Figure 2.8c shows another variation where the discharge occurs on a telephone set via the earphone and its loudspeaker capsule. The consequences in this case are severe: The ESD transient causes a cut-off of the conversation or loss of memorized phone numbers.

The current then returns to ground by all possible routes, with amplitudes prorated to the impedances of these respective paths. This means that the bulk of the current will flow by the lowest impedance path. Fig. 2.9 shows what these routes can be for a single stand-alone machine. In Fig. 2.9a, the path seems obvious since the machine is grounded by its safety wire and/or its neutral wire (since the neutral is generally grounded at the building level). However, this path does not resist a closer look. The free-space inductance of a round wire is about 1 μH/meter. A typical 2 meter cord would, therefore, have 2 μH of self-inductance. For a current rise time of 1 nanosec, this would give a dynamic impedance L/dt of $2\times 2.10^{-6}/1\times 10^{-9}=2000\ \Omega$, notwithstanding the additional length of building earth wire. Why would the ESD current run across more than 2 kilohms while a lower impedance path exists in parallel? Fig. 2.9b shows what this easier path can be.

Any machine containing conductive parts has a capacitance to ground. In the case of a metallic casing, the capacitance to ground can reach 100 or even 1000 pF. For instance, a mainframe having a bottom area of 1 m^2 and located 10 cm above ground will have a parasitic capacitance of 100 pF. For a rise time of 1 nanosec, this corresponds to a dynamic impedance of about 10 ohms!

If the machine is lifted .80 meters above ground on a non-

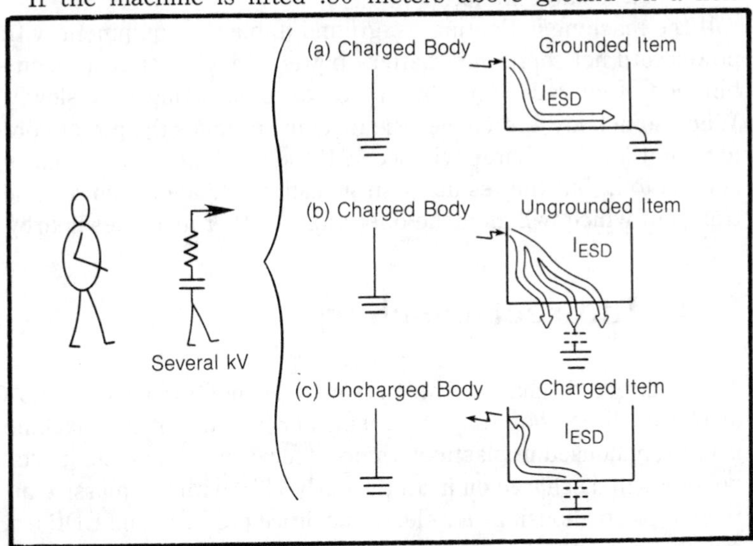

Figure 2.9—Personnel ESD coupling routes

conductive table, capacitance will decrease to 12 pF, representing about 80 ohms of impedance—still less than any green wire can offer. Therefore, a large proportion of the ESD current (especially during its rise, the most threatening one) will sink via the chassis-to-ground stray capacitance. This can be verified by a simple experiment: An ESD simulator is discharged on a grounded equipment. The power cord is then removed completely and the test is repeated. An arc will occur with no difficulty and, if a current probe is inserted over the generator tip, it will read about the same peak current. What will be affected is the *discharge* time constant of the current; in other words, the machine will stay charged up a longer time.

Does this mean that for a normally grounded machine, no current is flowing into the ground wire? Certainly not; the same current probe slipped over the power cord would read a current which can be a few or 10% of the total ESD current.

In Fig. 2.9c, a third possibility is shown—the reverse discharge. In this case, a machine has been charged by:

- successive previous discharges from people and objects
- internal static generations
- laminar flow of air, especially dry or cold air, or rubbing against a dry, isolated material

If the machine is floating vs. ground (table top equipment with power cord not connected, battery powered device, etc.), a recombination of charges is not occurring, or is occurring very slowly. When somebody approaches the machine or takes the power cord to plug it in, a discharge will occur. If the machine is off, no harm is done to it, but the resulting surge can create a local power line transient which will alter the operation of other machines nearby.

2.3 Indirect Discharge

In indirect discharge, the person does not (or even cannot) discharge directly on the equipment. For instance, if the machine is entirely housed in plastic with no or few accessible metal parts, nobody will discharge on it. In the early 70's, with the massive arrival of plastic housings for electronic office products and EDP terminals, there was a general belief that they would mark the end of the electrostatic nightmare. "Bye-bye ESD" was the song, but people did not dance to it for very long: field reports came in by legions to show that these products were experiencing even more ESD crashes than those with metal casings! Fig. 2.10 shows what happens. A person discharges on any nearby metallic part—a door frame, a pipe, furniture, maybe the very desk on which the machine is standing.* Then, the ESD pulse radiates a strong local electromagnetic field which couples into the nearby electronics since a plastic cabinet offers no shielding at all.

With the desktop unit, for instance (Fig. 2.10a), if the mother board lies flat on the bottom of the unit, the printed circuit is within 2 or 3 cm of the ESD current path.

Figure 2.10b is a more diabolic variation of indirect ESD witnessed by the author. A battery-operated calculator with a printer was being used by a draftsman. During the winter months, when the draftsman switched his desk-lamp on and off, the calculator would print a burst of figures. At first a power-line transient was suspected but, besides the fact that the unit was not plugged into anything, the printer would turn on even when the lamp was simply touched. The ESD current was flowing via the lamp shade and the stem, coupling to the power cord via the large capacitance between the flexible tube and the wire inside, then to the ground.

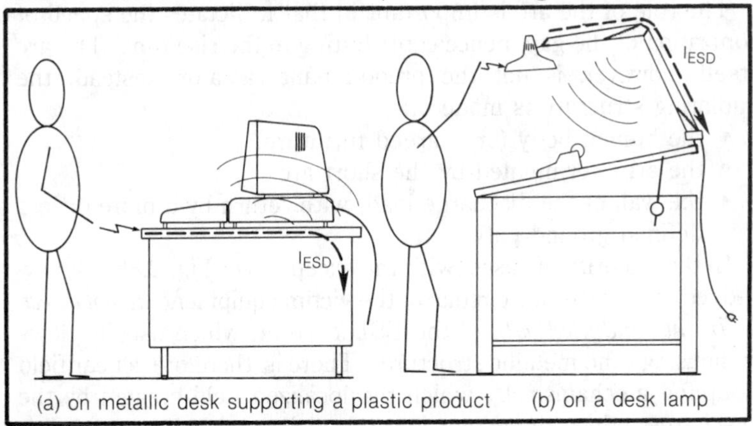

Figure 2.10—Indirect discharge

*The author remembers a case where, during a cocktail party, he was discharging himself on a large metal dish where the hors-d'oeuvres were arranged. The dish was on a cloth-covered wooden table, obviously grounded nowhere. But the charge transfer was enough to cause a violent ESD. Hopefully, sauteed chicken livers on toast and shrimps-a-la-creole are fairly ESD immune.

2.4 Coupling Mechanisms of ESD Transients into the Victim's Circuitry

All practical experiences have confirmed that the severity of ESD threats to machines is correlated to the magnitude of the current. This does not imply that it could not be predicted from the initial voltage, but everything being equal (generator capacitance, ESD loop dimension, initial voltage and, therefore, gap length), tests done with the least source resistance cause the most machine malfunctions.

The hypothesis that the arc length and the electric field in its vicinity are predominant factors in ESD interference is contradicted by this fact: A machine that withstands a 10 kV ESD with a generator having 1000 Ω of internal resistance will almost certainly fail at a lower level with a tester having only 150 Ω of internal resistance. In fact, the test condition considered the most severe of all is one where the arc no longer exists and the discharge is done with direct contact of the probe tip before the high voltage is applied.

The role of the arc is important in that it dictates the speed of ionization of the gap, hence contributing to the rise time. The arc itself, however, is not the predominant radiator. Instead, the radiating structure is made of:

- the human body (or charged furniture)
- the arm, terminated by the short arc
- the wall of the discharge load, with return by a more or less defined ground path

In the majority of cases (with the exception of Fig. 2.8b acknowledged), the electronic circuits of the victim equipment *are not directly in the conducted path* of the ESD current, which usually flows in housings and metallic structures. There is, therefore, a near field coupling mechanism by which the localized field created by the discharge induces a current into the pick-up circuit.

Fig. 2.11 shows a simplified explanation of this phenomenon based on the discharge current only. The first thing to point out

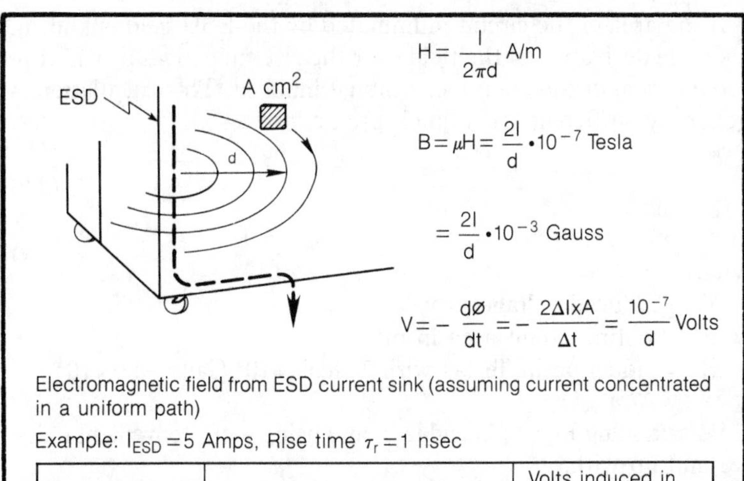

Electromagnetic field from ESD current sink (assuming current concentrated in a uniform path)

Example: $I_{ESD} = 5$ Amps, Rise time $\tau_r = 1$ nsec

Distance	H_{peak}	B_{peak}	Volts induced in 1 cm² loop
3 cm	24 A/m	.3 Gauss	3 volts
10 cm	8 A/m	.1 Gauss	1 volt
30 cm	2.4 A/m	.03 Gauss	.3 volt

Figure 2.11—ESD coupling by radiation

is that the dimensions of the ESD generating circuit are large compared to the distance to the receiving circuit. Therefore, it cannot be treated as a punctual source. Simple solutions of Maxwell's equation for small electric or magnetic doublets with their resulting $(1/d)^3$ and $(1/d)^2$ field dependency cannot be straightforwardly applied (a rigorous method would be to apply the method of moments to the current path broken down in small filaments). The model shown assimilates the ESD current path to a long wire model for which the resulting magnetic field is easy to calculate from the Biot and Savart law.

The ESD drain path to ground being long versus the distance of observation of the magnetic field is given by:

$$H = \frac{I}{2\pi d} \text{ Amp/m} \tag{2.1}$$

where,

I = ESD current in amperes
d = distance from ESD path to victim circuit.

If the area of the circuit illuminated by the ESD field is known, a simple derivation of the field over the rise time gives a crude approximation of the open loop voltage induced. This calculation is generally sufficient for a quick prediction.

$$V_i = -\frac{d\text{Ø}}{dt} = -\frac{dB}{dt}A \qquad (2.2)$$

where,

V_i = induced voltage in volts
A = victim circuit area in m^2
B = induction in Teslas with 1 Tesla$=10^4$ Gauss$=80\times10^4$ A/m

Re-arranging Eqs. (2.1) and (2.2) and using more convenient units, we end up with:

$$V_i = -\frac{2\Delta I\ A}{\Delta t \times d} \qquad (2.3)$$

where,

ΔI = change in ESD current in amperes
A = victim circuit area in cm^2
Δt = change in rise time in nanoseconds
d = distance from ESD path to victim circuit in cm.

Fig. 2.11 gives the results in voltages induced per cm^2 of victim area for 3 distances from the ESD path.

Consider this example: Two printed traces, one cm apart, having a 5 cm parallel run and located 10 cm from the ESD flow, will see a peak transient of 1 volt/$cm^2\times5$ $cm^2=5$ volts. This is more than enough to create an erroneous bit in most logic technologies.

Although very elementary, this method of predicting the ESD pulsed fields, and hence the induced parasitic voltages, gives adequate approximations when compared to actual measurement. A criticism could be raised against this overly simple model anyway: Considering the frequencies involved, the current flow should stay confined on the outer skin of the metallic cabinet (assuming a direct discharge) because of the skin effect. At 100 MHz, for instance, the skin depth in steel is about 30 microns; therefore, a 1 mm steel cover would be 30 skin depths thick and, according to basic shielding theory, no ESD current should be found on the inner side. This would be correct if the whole housing was a homogeneous shield, which is not the case by far—slots, joints, vents, displays, and cable entries create huge leakages, especially at these high frequencies. (This will be addressed in Secs. 5.4 and 5.5.)

The ESD current excites the multitude of slot antennas formed by box discontinuities, cooling apertures, ungasketed seams, etc. At the lower part of the spectrum, these leakage apertures represent a miniscule fraction of wavelength, and their attenuation is significant. However, for that part of the ESD spectrum which approaches or exceeds their $\lambda/2$ resonance, they *shine* inside with practically no attenuation. Now guess which part of the ESD radiated field induces the biggest voltages in the illuminated printed circuit boards and cables? The high frequency end, of course. Therefore, in an equipment not specially hardened against ESD (or high frequency EMI up to 1 GHz), the ESD current will flow on the inside of the cabinet as well as outside.

This radiated field from the ESD pulse, therefore, has two effects:

- It couples to the inner circuits of the machine: circuit boards, flat cables, discrete wiring,
- It exists outside (Fig. 2.12 shows external cables pick-up) all

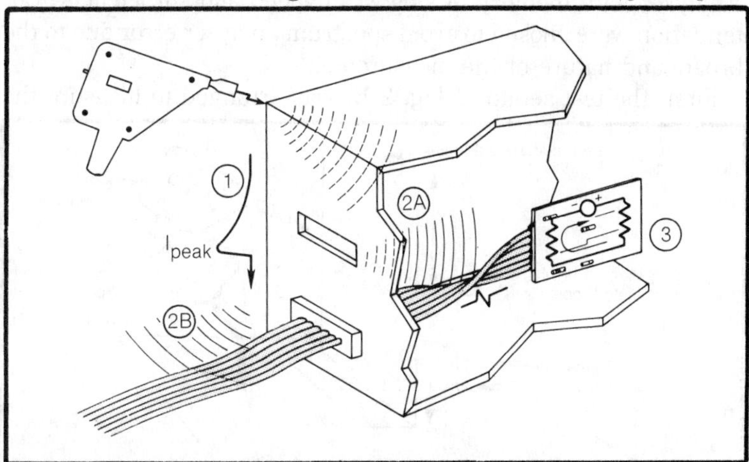

Figure 2.12—ESD coupling mechanisms. **(1)** The discharge current flows over the equipment cabinet. **(2A)** The cabinet imperfections let the high frequency components penetrate and shine inside (near and above the $\lambda/2$ of the seams. Also **(2B)**, the current radiates toward the I/O cables. **(3)** The circuits inside intercept the high frequency fields. Note that all these mechanisms are derivative; i.e., frequency dependent.

around the ESD source; thus, it couples into the external cables (signal and power).

Even though the current flows more like a spread stream than a thin wire, the H-field around its path can still be found by Eq. (2.1). However, knowing just the magnetic field may not be sufficient. A deeper knowledge of the nature of the electromagnetic field near the ESD path may be desired.

2.4.1 E and H-Field Values and Polarization from an ESD Charge to a Vertical Structure over a Conductive Ground

The author has conducted a set of experiments to evaluate both E and H-fields. In these tests, the electrostatic discharge was simulated using a Schaffner NSG 430 simulator (150 Ω, 150 pF network). The E and H-fields were measured by miniature monopoles, short balanced dipole, and magnetic loop (electrically shielded). The probes were connected to an Electro-Metrics ESA 1000 Spectrum Analyzer with memory. A slow scan speed and sufficient RF attenuation were chosen to avoid spectrum analyzer error due to the broadband nature of the measurement.

First, the test set-up of Fig. 2.13 was arranged to measure the

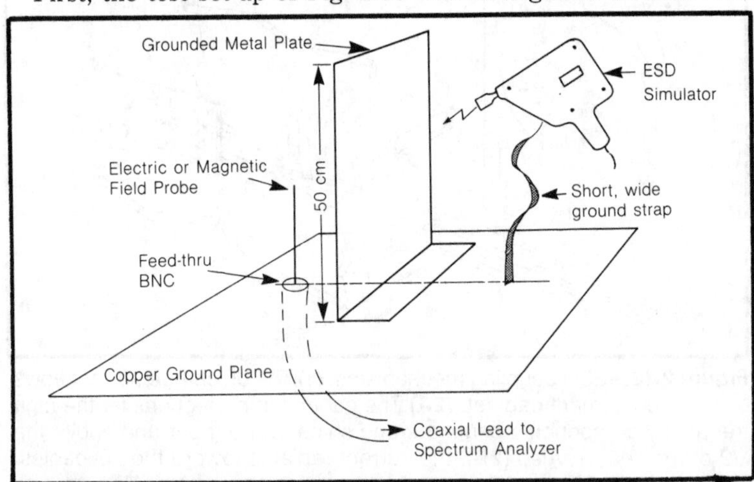

Figure 2.13—Experimental set-up for ESD field measurement

field amplitude facing an ESD "zap" in the absence of any protective shield. This would be the case of a discharge to a metal object near an equipment having only a plastic enclosure. This also serves as a reference for further shielding effectiveness assessment.

Although the non-uniformity of the field makes closer measurements less accurate, E and H-field magnitudes were measured at 10 cm, 30 cm and 1 meter during this study. The vertical structure was a 50 cm×6 cm aluminum plate firmly bonded to the copper ground plane. The ESD gun was set to 10 kV and an arc discharge with a slow repetition rate was made on the upper tip of the plate. The ground return for the ESD gun was a flat strap about 30 cm long to avoid the possible influence of both inductance and location of the return conductor. For the same repeatability reason, the orientation of this strap was always kept in the vertical plane formed by the gun and the structure which was discharged upon.

Figures 2.14 and 2.15 show the results of electric and magnetic fields after bandwidth and antenna factor correction. A few remarks are in order:

- The 1 meter results correlate within about ±15 dB with few other reported measurements done with similar discharge simulators (Refs. 16 and 21).
- Compared to the 1 meter results, the 30 cm and 10 cm results seem to show a $(\text{distance})^{-1/2}$ dependency instead of a $(\text{distance})^{-2}$ or $(\text{distance})^{-3}$ as one would expect, at least in the induction (near-field) region.

In this instance, it must be kept in mind that the radiator is the whole circuit formed by the simulator and the vertical discharge structure. Seen from an antenna located at ⩽1 meter, this structure behaves as an electrically long radiator, and not a punctual source or small doublet.

- Comparing E dBμV/m and H dBμA/m (see Table 2.2) shows a wave impedance varying, at a 1 meter distance, from about 100 Ω around 30 MHz to 300 Ω above 300 MHz. At 10 cm, the wave impedance varies from about 10 Ω around 10 MHz to 50 Ω above 300 MHz.

This indicates that for actual arcing on a metallic structure, the ESD generates a predominately magnetic (low impedance) field in the induction region, tending to a 120π ohms wave impedance in the far field zone. Since the change-over of near to far field is

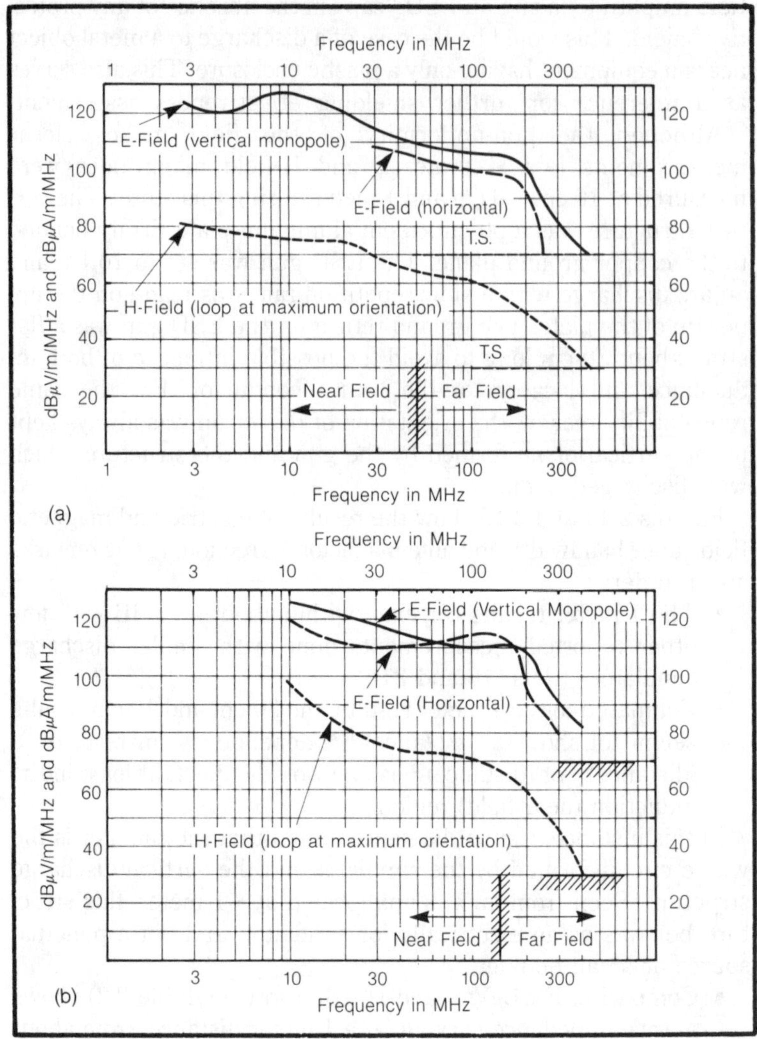

Figure 2.14—ESD fields at a distance of 1 meter (a) and 30 cm (b). The threshold of sensitivity T.S. is shown for reference.

wavelength dependent, the transition occurs at different frequencies for the various distances of the experiment; the change is very pronounced for the 10 cm case.

The fact that the field is predominately magnetic near the discharge path may be surprising. There is a common belief that

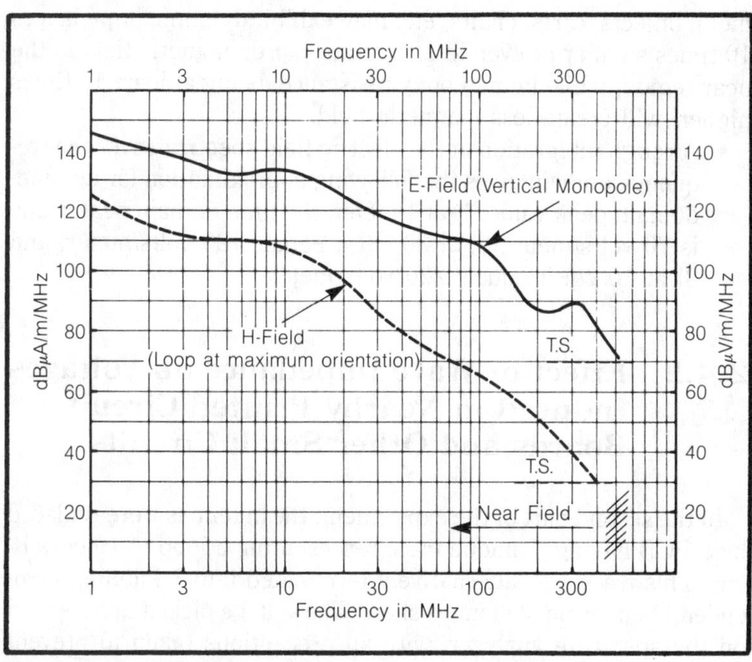

Figrue 2.15—ESD field at 10 cm

ESD, "being electrostatic, has to be an electric field." A close look at the discharge network can clarify this: Simple field theory says that in near field region, low impedance (<377 Ω) sources will radiate predominately magnetic fields, while high impedance (>377 Ω) sources radiate predominately electric fields. The ESD simulator used follows the IEC-65 recommendation and has an internal resistance of 150 Ω. Therefore, it behaves more like a magnetic source in the near-field. Will "real life" electrostatic discharges really appear like this? Actual furniture-type discharges from large

Table 2.2—Average wave impedance of ESD radiated fields

Frequency	3MHz	30MHz	100MHz	300MHz	500MHz
Distance 1m	Near Region	Transition		Far Region	
$Z_{wave\ Aver}$	100Ω	120Ω	160Ω	317Ω	
Distance 10 cm	Near Region				Trans. Region
$Z_{wave\ Aver}$	14Ω	40Ω	158Ω	178Ω	

metal objects, carts, chairs, etc. may exhibit dynamic impedances 10 times smaller or even less, creating more magnetic field in the near region, while human body resistance, being at least 10 times higher, will create less magnetic field.

- A rough integration of the electric field spectrum over the frequency domain gives the following approximation for its time-domain peak value: 1) at 1 meter, the time-domain peak value is 70 volts/meter; 2) at 30 cm it equals 120 volts/meter; and 3) at 10 cm it equals 220 volts/meter

2.4.2 Effect of Wave Impedance on Voltages Induced in Nearby Printed Circuit Boards and Other Small Circuits

In the second part of the experiment, the antennas were replaced by a PCB having a unique trace representing a loop of 10 cm×10 cm. This trace was alternatively terminated into 1 kilohm, open-ended, then terminated into a short. The voltage picked up was read on the spectrum analyzer, with all precautions taken to prevent possible pick-up by the coaxial cable. The effects of varying the far and terminating resistances is interesting in the perspective of understanding whether H-field or E-field coupling predominates.

Fig. 2.16 shows the induced voltages in dBμV/MHz, the PCB being oriented tangent to the wave front.

Fig. 2.17 shows the traditional model for a small rectangular circuit illuminated by an electromagnetic field. The E-field creates a transverse voltage V_2, which appears as a high impedance source (current source) with an open circuit voltage:

$$V_2 = E \times 2\ell \times h \times \frac{\pi}{\lambda} \cos\theta \cos\alpha \tag{2.4}$$

where

θ = angle between the E-field and the direction of ℓ

α = angle between the plane of the loop and the direction of propagation.

Given the 50 Ω impedance of the "victim" receptor near end (spectrum analyzer), a high impedance on the far end, such as 1 kΩ or ∞, will double the available transverse voltage while a shorted end will nullify it.

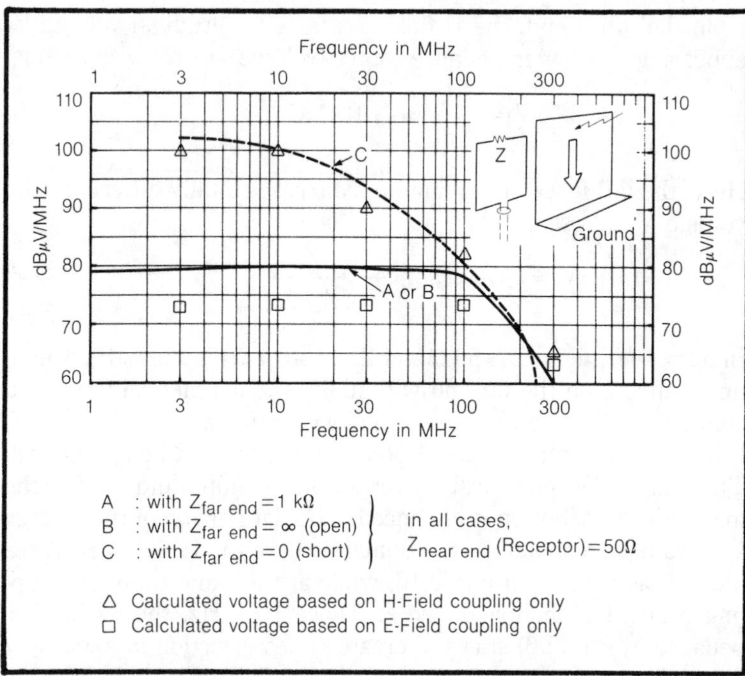

Figure 2.16—Broadband voltage induced in a 100 cm² PCB run located 10 cm from the ESD path, parallel to wave front (PCB not oriented for maximum H-field interception)

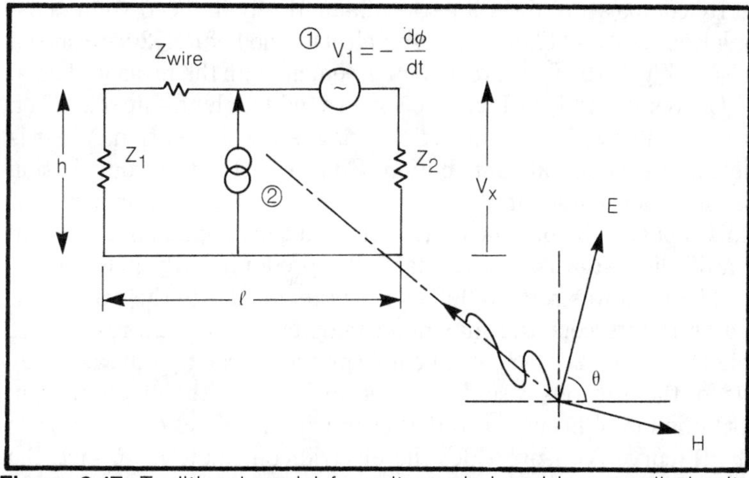

Figure 2.17—Traditional model for voltages induced in a small circuit illuminated by an EM field

On the other end, the H-field creates a longitudinal voltage V_1 appearing as a low impedance source (voltage source) with a value:

$$V_1 = -\frac{d\varnothing}{dt} = -\omega B \times \ell \times h \qquad (2.5)$$

Here, the influence of varying impedances is totally different. The voltage V_x across the receptor end is:

$$V_x = V_{1\ (H\ coupling)} \frac{Z_2}{Z_1 + Z_2 + Z_{wiring}} \qquad (2.6)$$

In our experiment, Z_2=50 Ω and $Z_{wiring} \cong 0.1\ \Omega + j\omega \times 0.4\ \mu H$. A high impedance Z on the far end will nullify the magnetically induced voltage. A *short on the far end will maximize it.*

In Fig. 2.16, curve C corresponds to the far end being shorted. Therefore, the electrical contribution is minimum while the magnetic contribution is enhanced. The flat portion of the voltage spectrum corresponds to the domain where the H-field decreases like 1/F as evident in Fig. 2.16, while at the same time the coupling coefficient increases like F. Above 10 MHz, the wiring impedance in Eq. (2.6) starts to create series insertion loss, causing less and less voltage available at the 50 Ω end. Finally, above 150-200 MHz, the H field spectrum itself decreases like $1/F^2$, causing the available voltage to collapse even more rapidly.

By comparison, the values for magnetically induced voltage using Eqs. (2.5) and (2.6), or the graphical method of Ref. 20 are shown also in Fig. 2.16. They are in fair agreement with the measured data.

Curves A and B of Fig. 2.16 correspond to a high impedance on the far end, which minimizes the magnetic contribution. What is left is the electrical contribution. It is clear that this contribution is one order of magnitude less then the magnetic one. Furthermore, this supports the previous statement that the radiation of the "standard" discharge is, in the near field, predominantly magnetic.

The PCB was rotated 90° to be perpendicular to the wave front so as to intercept the maximum magnetic flux. The results are shown on Fig. 2.18 and can be interpreted in exactly the same way as for the previous case. In this set-up, however, the magnetic contribution is so pronounced that even with a far end termination of 1 kΩ (curve A) it overrides the electric contribution. In fact, the difference between curves A and C below 10 MHz correspond approximately to the ratio of 1000 Ω to 50 Ω termination. Interestingly enough, integrating the spectrum of curve C corresponds to a peak voltage of about 15 volts.

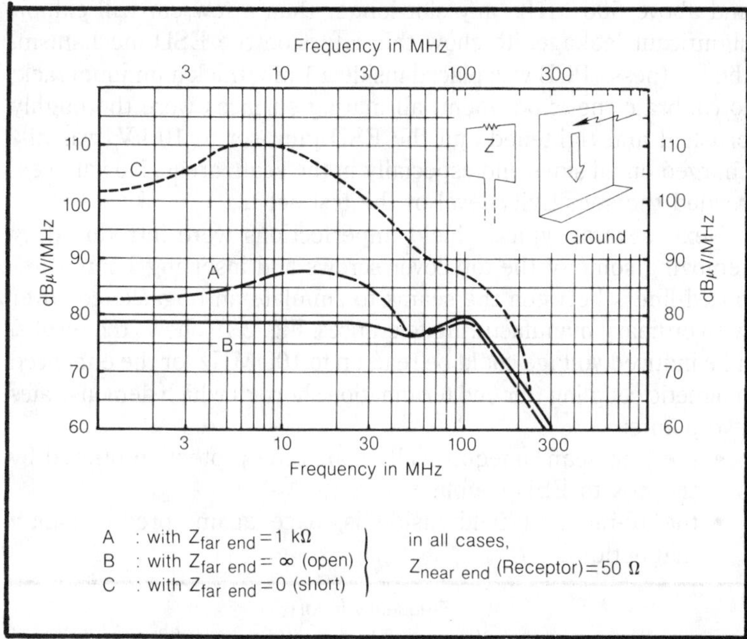

Figure 2.18—Broadband voltage induced in a 100 cm² PCB run located 10 cm from the ESD path perpendicular to wave front (PCB intercepting maximum magnetic field)

2.4.3 Effect of a Typical Metallic Cabinet on ESD Radiation

If, instead of being plastic, the cabinet housing the electronic circuit is metallic (or metallized), the ESD field should be attenuated by the normal shielding effect of the material. Therefore, and not accounting for the external pick-up by I/O cables which was shown in Fig. 2.12, it seems that internal circuit boards and wiring should be fairly well protected. However, real life enclosures are full of slots, seams, apertures, etc., which disrupt the shield integrity.

Every shield discontinuity across the ESD current path will "shine" inside with an efficiency proportional to its length compared to the half wavelength. The ESD spectrum extending up to and above 500 MHz, any slot longer than a few cm will exhibit significant leakage. To show this effect on the ESD mechanism, the "witness" PCB was placed inside a 1 mm thick aluminum rack; to calibrate the experiment, all mating surfaces were thoroughly brushed and tightened and the ESD gun, set to 10 kV, was discharged on all sides and especially in the seam areas. No value exceeded the sensitivity level of the test set up.

Next, several typical shield imperfections were introduced by removing some of the top cover screws and inserting 1 mm cardboard liners between the seams to simulate an ungasketed cover with ordinary manufacturing tolerances. Fig. 2.19 shows the results. The induced voltage could be read up to 100 MHz for the enhanced magnetic coupling (far end termination shorted) which demonstrates two points:

- the thin seam unequivocally spoils the protection offered by the box to ESD coupling
- the re-radiated field inside is, once again, predominantly magnetic

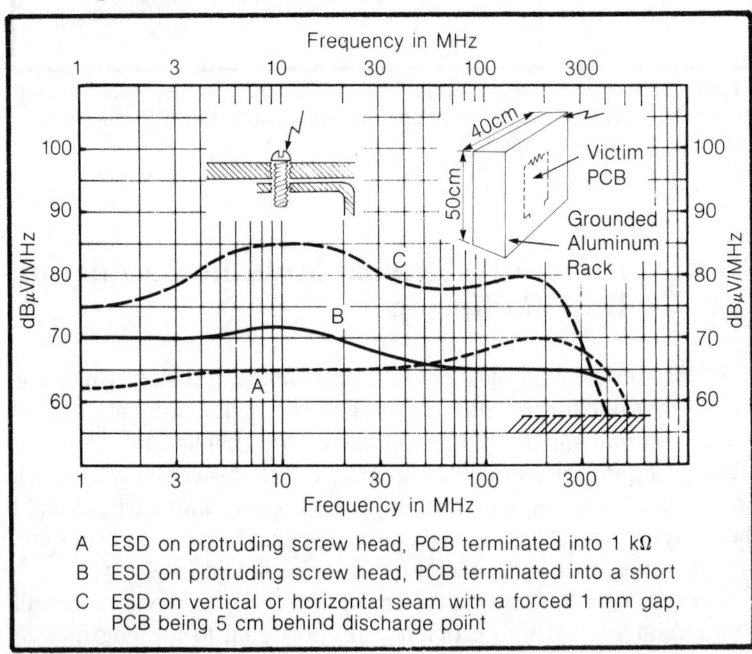

Figure 2.19—Voltage induced on the 10 cm×10 cm PCB trace housed in an aluminum rack

An interesting effect was also simulated: One of the threaded holes used to attach the top cover was painted and the cover was mounted using a long screw protruding about 2.5 cm (1 inch) inside the box. The results are also shown in Fig. 2.19. It seems that the screw generates a secondary arc inside, between the fillet and the inner box surface.

Finally, another coupling mechanism, generally of the second order, is common impedance pollution. (See Fig. 2.20.) Even if they are conductive, the metal or metallized covers have a certain RF impedance. The impedance of a 1 mm thick steel plate for a 1 nsec rise time is 0.4 ohms/square. The impedance of a graphite paint used to shield plastic housing is 10 ohms/square. For a 10 amp ESD current, this will correspond to an instantaneous ground shift of 4 volts and 100 volts, respectively, between points A and B in Fig. 2.20. The two electronic subassemblies grounded at these points will experience this common-mode voltage.

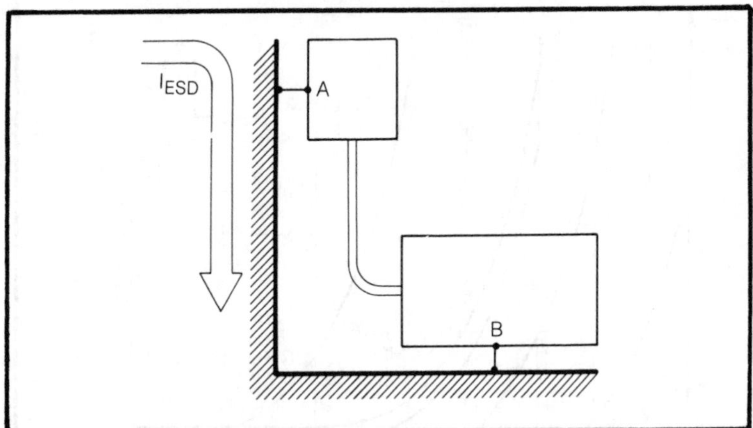

Figure 2.20—Common impedance coupling with ESD current

2.5 Response of Victim Circuits and Types of Errors

So far, we have evaluated the peak amplitude of voltages induced into victim circuits illuminated by the ESD pulse. In many cases, the amplitudes were far above the logic thresholds, so an erroneous change in logic status was expected. However, the induced transient is extremely short and it must be compared, on a time scale, to the ac noise immunity of the victim logic (Fig. 2.21).

For instance, a standard TTL gate with 10 nsec speed has a worst-case dc noise margin of 0.4 volts (0.8 V typical) for a junction temperature of 85°C. But, if the noise pulse is shorter than 10 nsec, it takes more amplitude to trigger the gate; that is, to provide the amount of nanojoules the gate needs to operate.

As shown in the figure, the faster the logic, the more vulnerable it will be to ESD *induced* pulses. One could also compare the frequency spectrum of ESD with the various bandwidths of CMOS, TTL, ECL, etc. (Fig. 2.22).

In any case, it is important to take into consideration the coupling factor by which the ESD transient induces a parasitic voltage. The coupling factor is a frequency dependent term; the induced

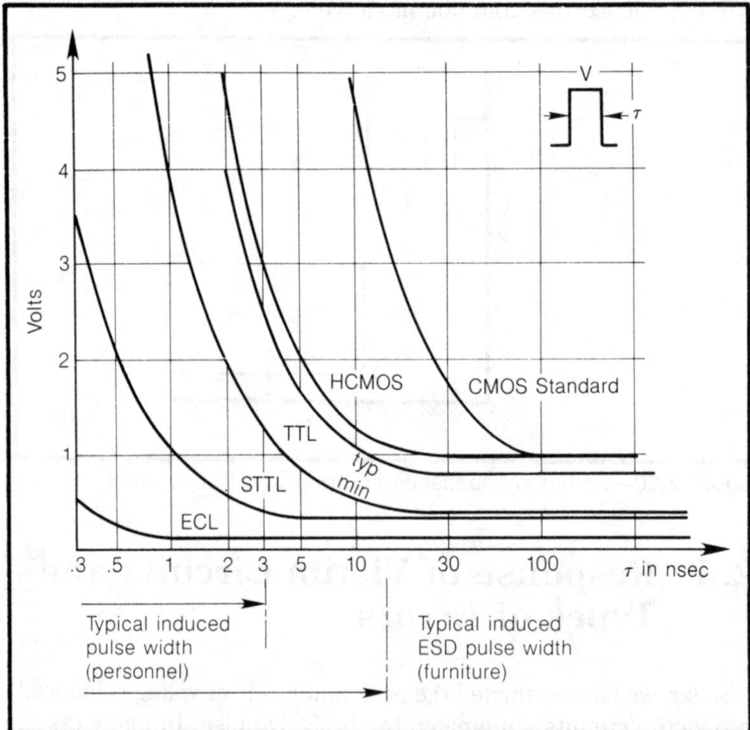

Figure 2.21—Noise margin of current logics versus pulse width

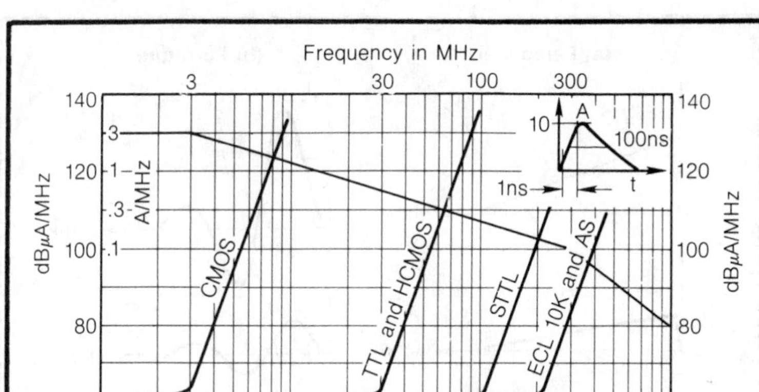

Figure 2.22—Time and frequency aspect of a 10 kV personnel ESD current similar to the typical waveform of Fig. 1.18, shown in contrast with some typical logics bandwidth

voltage is not a replica of the ESD waveform, but its derivative. Thus, the duration of the induced spike will be equal to the rise time of the ESD current (the time during which the derivative dØ/dt exists).

This is shown in Fig. 2.23a. Note also the sign of the voltage V_i, which is the opposite of the sign of the field change. Of course, in terms of the actual polarity of the induced noise, the sign of the voltage will depend on the orientation of the victim circuit. In Sec. 1.3, we did say that, depending on the nature of the triboelectric materials which created the electrostatic charge, the ESD pulse can be positive or negative. A polarity reversal in the ESD pulse would create a corresponding reversal in the induced noise. A typical logic gate is susceptible to a positive going level when it is in a low input (Logic "O") state, and it is susceptible to a negative going level when it is in a high input (Logic "1") state. *Therefore, it would not be irrelevant to perform an ESD test with both polarities;* depending on what the logic circuits are doing at the time of the ESD event, the equipment under test could be more vulnerable to one polarity than the other. Fig. 2.23b shows the same cascade of couplings for a furniture discharge. Since the rate of charge of the ESD current in amps/nanosec is generally less than for personnel, it seems that

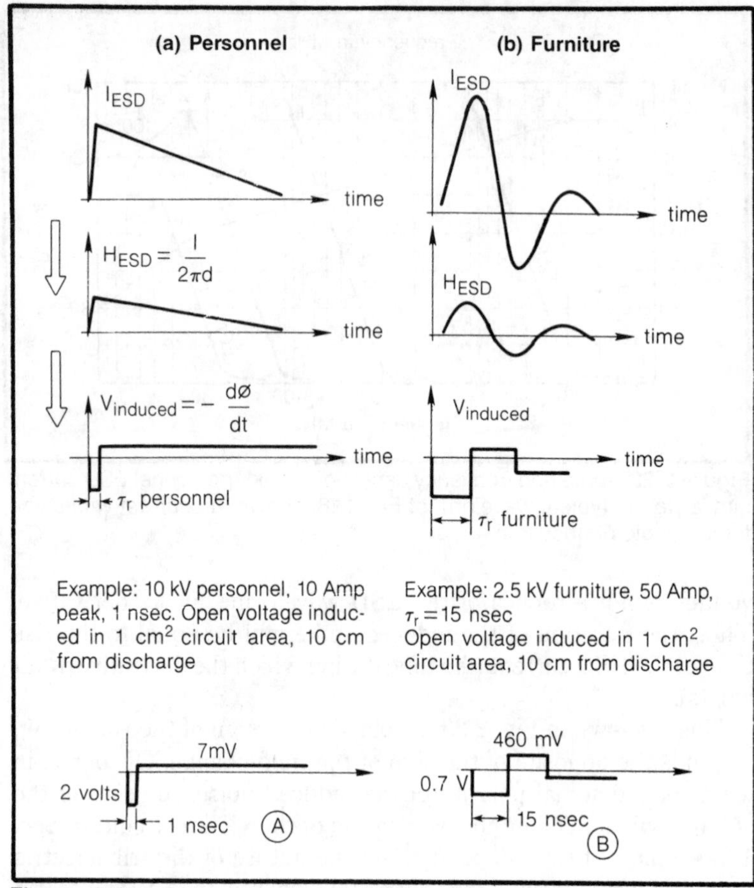

Figure 2.23—Comparison of induced voltages from personnel and furniture ESD, showing influence of time derivation

the threat is less severe. *However, since the rise time is much longer* (15 nsec, for instance), *the induced spike is also longer and will fall into the dc noise margin of most logics.* This is why, among other things, a furniture discharge can be a more threatening kind of ESD.

2.5.1 Influence of Circuit Impedances

In most cases where the ESD event is causing a high peak current, we have seen that the strongest contributor of induced noise is the magnetic field. Consequently, the induced voltage will appear in series in the circuit. The resulting voltage at the victim component input will be:

$$V_{diff} = V_i \frac{Z_L}{Z_L + Z_s + Z_w} \tag{2.7}$$

where,

Z_L = input impedance of the victim component (for example, logic gate, operational amplifier, etc.)
Z_S = source impedance of the driver
Z_W = impedance of the wiring

Since, for obvious reasons, the victim input generally has a higher

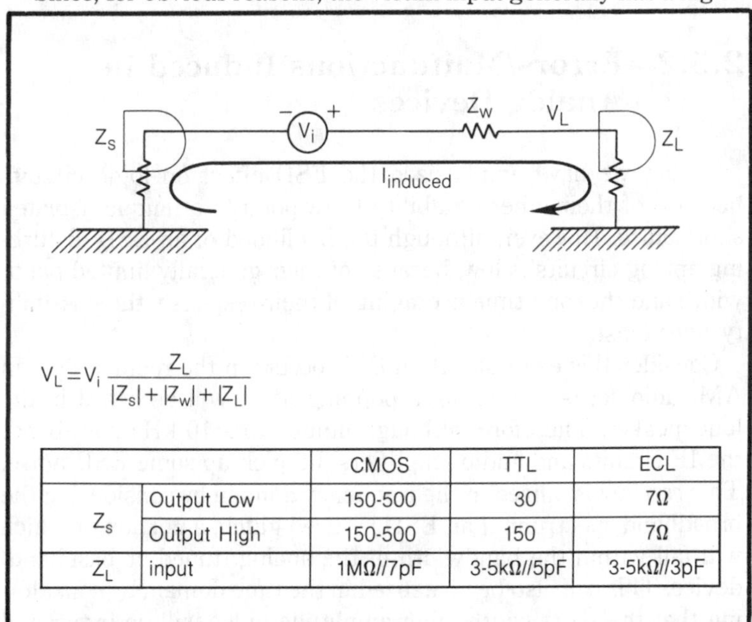

		CMOS	TTL	ECL
Z_S	Output Low	150-500	30	7Ω
	Output High	150-500	150	7Ω
Z_L	input	1MΩ//7pF	3-5kΩ//5pF	3-5kΩ//3pF

Figure 2.24—Influence of victim circuit impedance on coupled voltage

impedance than the signal source, Eq. 2.7 shows that most of the noise will appear at the victim input. For the same reason, this usual impedance configuration will cause the noise coupled by the electric field to be less of a problem, since it will appear differentially across the line and see the parallel combination of Z_L shunted by Z_S.

In any case, two other considerations may prevail:

- above a certain level (30-50 V for certain technologies), even if the device is not triggered, some permanent overstress may have occurred
- certain technologies, like CMOS, exhibit a latch-up mechanism if the input becomes more positive than V_{cc} or becomes more negative than the 0 V reference. In these conditions, the parasitic PNP and NPN transistors within the device behave like an SCR which is fired. Even after the pulse has gone, the SCR stays in conduction. The current is limited only by the external loads and can lead the device to destruction, since the parasitic SCR can only be turned off by powering off the supply.

2.5.2 Errors/Malfunctions Induced in Analog Devices

So far, we have emphasized the ESD effect on logic circuits because of their inherent ability to respond to a unique, isolated short pulse. However, although the likelihood of an ESD disturbing analog circuits is low, because of their generally limited bandwidth and the long time averaging of their response, the possibility does exist.

Consider this example: If an ESD occurs in the room where an AM radio set is turned on, a popping noise will be heard in the loudspeaker. Therefore, although limited to a 10 kHz pass-band, the IF circuits and audio-amplifiers did pick up some ESD noise. This can be visualized in the frequency domain by considering the broadband spectrum of an ESD induced glitch, a portion of which will fall within the bandwidth of the analog (tuned or baseband) device. This can also be visualized in the time domain by considering that the short width, high amplitude pulse will be integrated through the long time constant of the analog device. The pulse will become longer with less amplitude, though enough to initiate a circuit response, since the threshold of analog amplifiers is usually very low. Figure 2.25 shows this concept.

Besides this dynamic aspect, low speed analog devices can present another inherent problem since they generally have a very high input impedance. Thus, in the presence of an electrostatic field (without a discharge), the input(s) can rise to any voltage, depending on the divider network formed by the elctrostatic source to device capacitance and the device to ground capacitance. Solutions to this problem (not actually an ESD) are presented in Chapter 5.

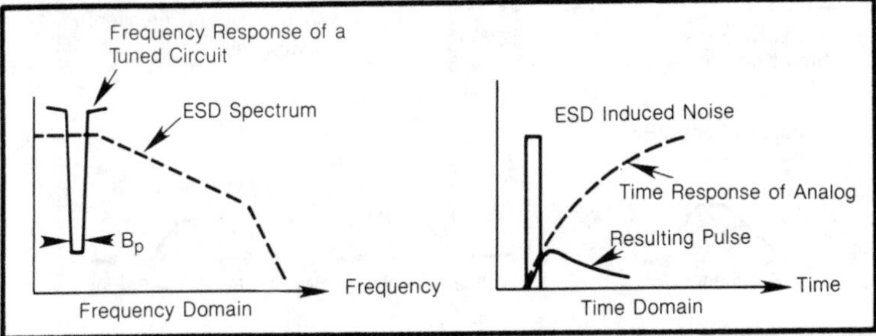

Figure 2.25—Concept of the response to ESD noise of an analog device

2.6 Prediction of ESD Induced Noise by Frequency Domain Estimations

The previous examples essentially showed the coupling mechanisms using time domain calculations. For those more used to frequency domain, the following examples will show a prediction of the ESD induced noise by using frequency spectrum considerations.

The principle of the frequency approach, very classical in EMI (for instance, see Ref. 20), can be summarized in Fig. 2.26. Time domain has the advantage of giving straightforward solutions that are rather accurate. In contrast, frequency domain requires some more complex transformations. In addition, the phase (or polarity) information is lost. On the other hand, when there is a cascade of complex mechanisms like induction+$\lambda/2$+resonance+shielding+bandwidth rejection, etc., since these parameters are generally better described in frequency domain, a more organized approach to the solution can be made.

Two sets of curves have been developed to speed up the prediction process. The graph of Fig. 2.27 allows prediction of the magnetic field at a given distance "d" from a long conductor (or path) carrying a current I (Ampere's law for a long wire). The graph

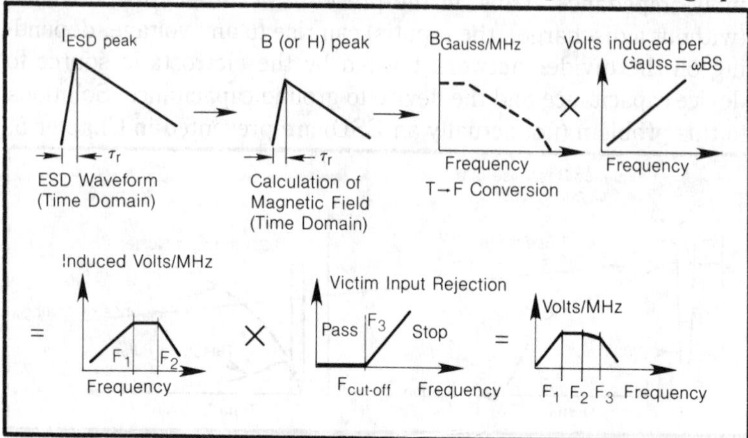

Figure 2.26—Frequency domain prediction of ESD coupling

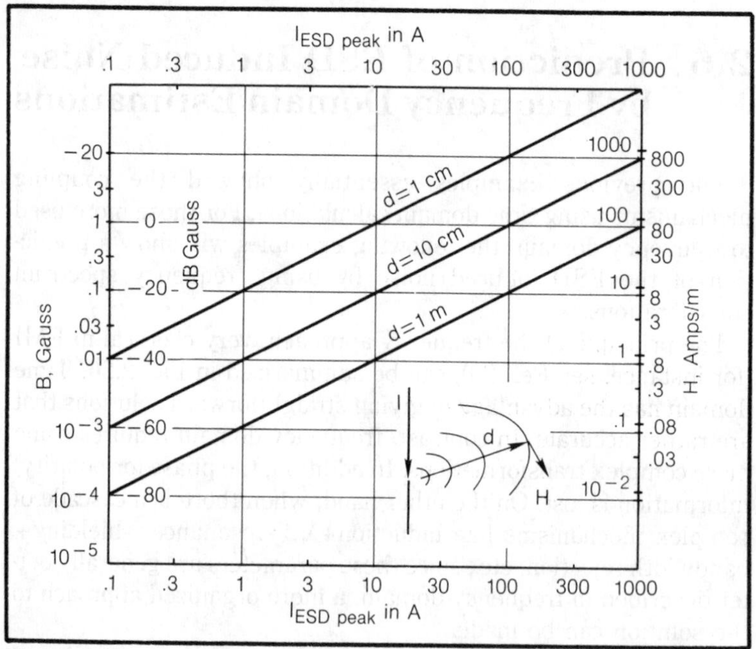

Figure 2.27—Magnetic field at a distance d from a current path I

of Fig. 2.28 gives the induced voltage in volts per Gauss of ambient field in a given circuit area $\ell \times s$. All curves increase with frequency (+20 dB per decade) until the longest element of the pick-up circuit reaches $\lambda/2$, where it clamps to the maximum corresponding to a half-wave dipole. To explain the method, we will run a numerical example using two methods—a fast approximation and a more precise graphical method.

The ESD scenario is shown in Fig. 2.29. A personnel ESD occurs near a flat cable running 10 cm from the ESD drain path.

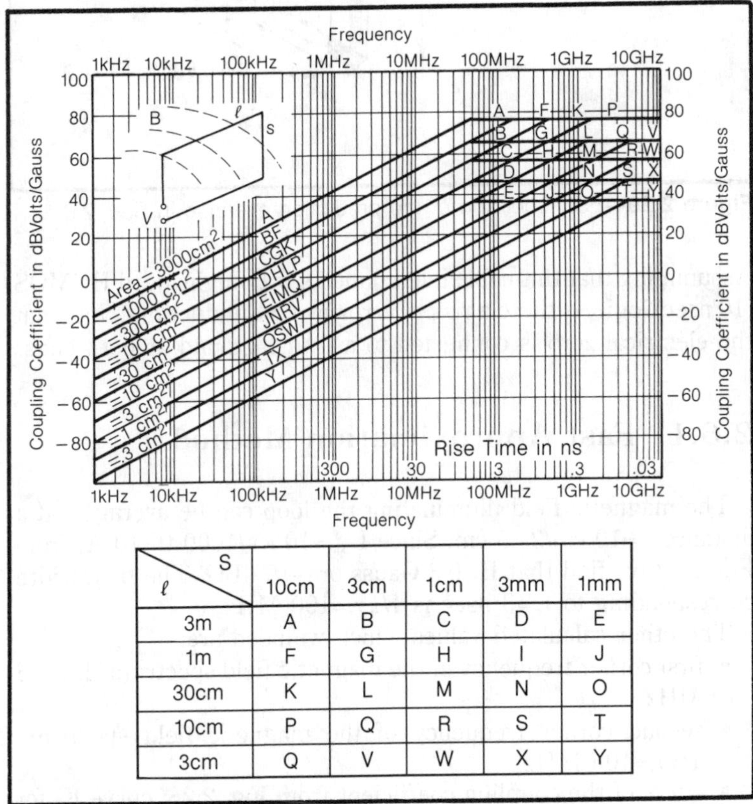

S / ℓ	10cm	3cm	1cm	3mm	1mm
3m	A	B	C	D	E
1m	F	G	H	I	J
30cm	K	L	M	N	O
10cm	P	Q	R	S	T
3cm	Q	V	W	X	Y

Figure 2.28—Induced voltage vs. area and frequency

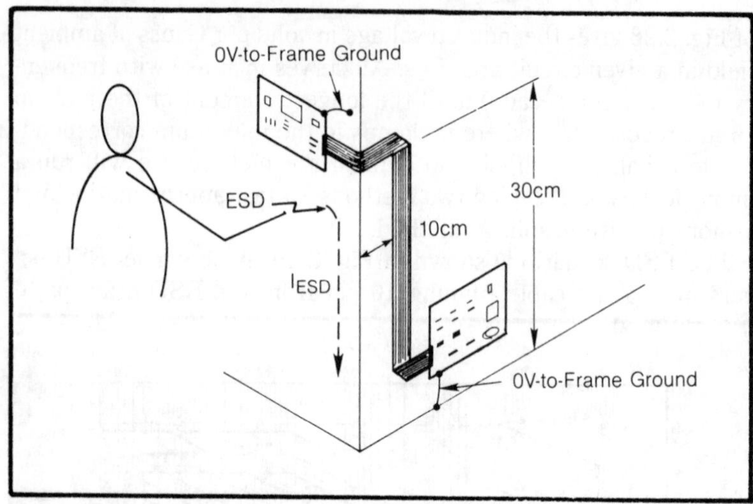

Figure 2.29—ESD scenario of the prediction exercise in Sec. 2.6

Assumming that this cable interconnects with high-speed HCMOS (10 nsec) logic, is there a risk of error by ESD induction? Note that the electrical zero is connected to a frame ground at both ends.

2.6.1 Fast Approximation Method

The magnetic field illuminating the loop can be averaged at a distance d=10 cm/2=5 cm. Since I_{ESD}=10 kV/1000 Ω=10 A, from Fig. 2.27 we find that B=0.3 Gauss or −10 dBG. The bandwidth corresponding to τ_r=3 nsec is $1/\pi\tau_r$=100 MHz.

The other calculated values which we need are:

- first corner frequency of the magnetic field spectrum: $1/\pi\tau$=3 MHz
- second corner frequency of the magnetic field spectrum: $1/\pi\tau_r$=106 MHz
- curve of the coupling coefficient from Fig. 2.28: curve K, for 30 cmx10 cm
- bandwidth of HCMOS: $B_p=1/\pi\tau_r=1/\pi x10^{-8}$=30 MHz

In Fig. 2.28, we localize the pick-up loop of 30 cm×10 cm as curve K.

The coupling coefficient on curve K for 100 MHz is 62 dBVolts/Gauss. So the induced voltage is

$$V=62 \text{ dB V/G}+(-10 \text{ dB G})=52 \text{ dBVolt}\cong 350 \text{ volts.}$$

This voltage is a common-mode voltage appearing in series in the loop.

Since the PCB's zero volts are grounded at both ends, the rejection of CM noise is extremely poor. In addition, since the CMOS is a high input impedance, virtually the full CM voltage will appear at the input of the victim circuit. However, the HCMOS has less bandwidth than the spectral occupancy of the 3 nsec ESD pulse. Therefore, to compare with the in-band threshold, an additional rejection must be accounted for:

$$\text{BW Rejection}=\frac{\text{HCMOS Bandwidth}}{\text{ESD Bandwidth}}=\frac{1/\pi\tau_r \text{ (CMOS)}}{1/\pi\tau_r \text{ (ESD)}} \tag{2.8}$$

$$=\frac{\tau_r \text{ (ESD)}}{\tau_r \text{ (HCMOS)}}=\frac{3 \text{ nsec}}{10 \text{ nsec}}=0.3$$

Thus, the equivalent input voltage will be $350\times0.3=105$ V.

In conclusion,

- the 350 V transient represents a risk of latent damage
- at 105 V, the equivalent input voltage is 100 times the dc noise margin!

2.6.2 Graphical Method

A more elaborate graphical method will now be explained. We will start by calculating the magnetic field spectral density. From the calculation in Sec. 2.6.1, induction B=0.3 Gauss, $\tau=0.69\times RC=100$ ns, and $\tau_r=3$ nsec. The spectral density at the pedestal=$2\times B\times\tau$ μsec=$2\times0.3\times0.1=0.06$ dB Gauss/MHz or −24 dBG/MHz.

The graphical construction of victim voltage is shown in Fig. 2.30:

- The mid-band amplitude for segment A at 1 MHz is +8 dBV/MHz+20 $\log_{10}$ 1 MHz/3 MHz=−2 dBV/MHz or 0.8 V/MHz.

- The mid-band amplitude for segment B is +8 dBV/MHz or 2.5 V/MHz.
- The mid-band amplitude for segment C at $F_c = \sqrt{30 \times 100} = 54$ MHz is +8 dBV/MHz−20 $\log_{10}$ 54/30=+3dBV/MHz or 1.4 V/MHz.
- The mid-band amplitude for segment D at $F_c = \sqrt{100 \times 500} = 220$ MHz is −2 dBV/MHz−40 $\log_{10}$ 220/100=−15.6 dBV/MHz or .16 V/MHz.

Beyond 500 MHz, the voltage amplitude falls off at 60 dB/decade; i.e., like $1/F^3$, and is neglected.

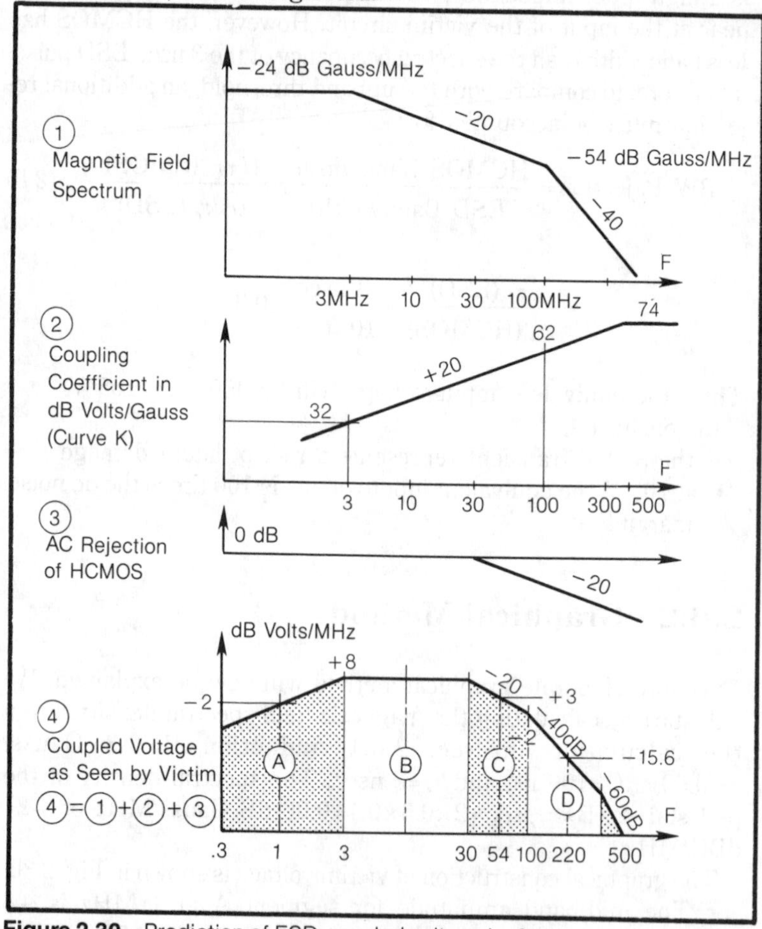

Figure 2.30—Prediction of ESD coupled voltage by frequency plot method

Table 2.3—Total induced voltage calculations

Segment	1 Amplitude (Volts/MHz)	2 Frequency Interval (MHz)	Resulting Voltage 3=2×1
A	.8	3−0.3=2.7	2.1
B	2.5	30−3=27	67.5
C	1.3	100−30=70	91.0
D	.16	500−100=400	64.0
		Total	224 Volts*

*Since these spectral segments are certainly not in phase, it would probably be more correct to make their RMS addition, i.e., $V_{TOT} = \sqrt{A^2+B^2+C^2}$ etc.

What remains is to integrate each segment over its respective frequency span. This is shown in Table 2.3. By comparing these results with the results of the fast approximation method of 105 volts, we see that it was optimistic by about +100% or 6 dB, which is certainly satisfactory enough. But what is interesting about the graphical method is that it shows in which frequency areas the solutions should be optimized. For instance, floating the PCB 0-volt at one end, normally a good solution for avoiding ground loops, would not be very efficient here. The parasitic capacitance of a PCB to chassis is usually in the range of 30 to 100 pF; i.e., the isolation from chassis vanishes above approximately 10 MHz, while the biggest contribution is made by the portion of the spectrum from 3 to 500 MHz. In contrast, ferrites and ceramic decoupling capacitors would work at their best. All these solutions will be discussed in Chapter 5.

Incidentally, approaching the cable from the frame in an attempt to reduce the pick-up loop size would not help because, at the same time, the average distance from the current path would also decrease, hence increasing the field intensity.

2.7 Personnel ESD or Furniture ESD: Which One is Worse?

From what we have seen, it appears that furniture ESD, due to its higher current and resultant magnetic field, is the bigger threat.

However, consider this:

- Furniture ESD will only occur at some points of the victim equipment. Some other areas will never be touched by a moving cart, chair, etc.
- Furniture ESD usually exhibits much lower voltages than personnel. Therefore, recessed parts and metal parts or circuits behind a plastic barrier will never be zapped by a furniture ESD because the voltage does not reach the breakdown voltage of the air gap or plastic barrier. In contrast, a 15 or 20 kV personnel ESD will reach the breakdown voltage by arcing these remote points.

Therefore, as said in Chapter 1, it seems that a complete ESD simulation should address both type of discharges.

Chapter 3

ESD Specifications

Since the discovery of the ESD threat to electronic equipment, several ESD testing specifications have been established by international or professional bodies. Although they all claim to characterize the same phenomena, the values recommended by the different organizations vary widely depending on the type of application they deal with. Some of them specifically call for a maximum rise time (or a minimum dV/dt or dI/dt), while others do not. Table 3.1 gives a recapitulation of the various ESD test values recommended. The voltage is labeled "maximum" because many specifications regulate amplitude as a function of the type of environment (controlled or not) in which the product is marketed.

ESD specifications can be classified according to two simple categories:

- those which specify the devices' (integrated circuits and discrete components) immunity to static discharges to determine the risk of immediate or latent failures
- those which specify the immunity of electronic equipment to ESD; i.e., ESD induced errors or alterations.

3.1 ESD Specification for Device Sensitivity: MIL-STD-883 and 1686

Military Standard 883 (1983 revision) is a voluminous document describing all reliability tests (electrical, mechanical, etc.) for micro-

Table 3.1—Summary of Current and Proposed ESD Standards

Category 1 (For Devices)	Max. Voltage	Polarity	Discharge Network	Comments
MIL-STD-883	2000 V (Class 1 sensitivity)	+/−	100pF/1500Ω	5 pulses waveform specified
DOD-STD-1686 & HDBOOK 263	4000 V (Class 2 sensitivity)	+/−	100pF/1500Ω	
	4000-15000 V (non sensitive devices)	+/−	100pF/1500Ω	
STACK (European manufacturers)	500V	+/−	100pF/1500Ω	5 pulses
Category 2 (For Equipment)				
IEC-801-2	15,000 V	+	150pF/150Ω	10 pulses waveform specified
SAE, Committee AE-4 (Computers)	12,000 V (controlled environment)	+	100pF/100Ω	50 pulses
	12000 V (uncontrolled environment)	+	10nF/10Ω	50 pulses
SAE-J 1113/B (Automobile)	20,000 V	+	300pF/5000Ω	
EIA 1361	10,000 V	+/−	100pF/500Ω	
NEMA, Part DC 33	20,000 V	+/−	100pF/1500Ω	
3M (Internal Specification)	10,000 V	+	500pF/100Ω	
IBM (Internal Specification)	Not Published	+	150pF/2000Ω 150pF/15Ω	personnel furniture

circuits. Electrostatic discharge is covered only in a tiny portion of the whole standard which is known as Method 3015.2. The highlights are the following:

- According to the results, the devices will be classified into Category A(1)—sensitive to 2000 V or less, or Category B(2)—sensitive to more than 2000 V.
- The pulse waveform is defined across a 1500 ohm load substituted to the DUT. The maximum rise time is 15 nsec.
- The pulse is applied (+) and (−) between each pin and the device common, with 5 pulses each time.

MIL-STD-1686 (1980), issued by the U.S. Department of Defense, is deemed to be a complementary document dealing uniquely with

ESD. Curiously, however, the two standards do not cross-reference each other which leaves some doubt as to whether or not they are part of a concerted strategy. Specifically, although the test circuits are the same, the pulse amplitude and waveform is not defined in MIL-STD-1686 as it is in MIL-STD-883. In any case, MIL-STD-1686 is a very complete document including ESD recommendations and control plan (see Sec. 2.1). The test portion can be summarized this way:

- Semiconductors which are sensitive to 4000 V or less must be characterized according to a two-class system:
  class 1: ⩽ 1000 V sensitivity
  class 2: > 1000 V and up to 4000 V sensitivity.

  In contrast to MIL-STD-883, no pulse waveform is specified.
- The test requirements are:
  a. Perform an electrical characterization (dc) on each pin of the device *before* testing.
  b. Apply the ESD pulses (no minimum number specified)
     (1) All pins together against center of the integrated circuit case, (+) and (−)
     (2) Each pin against common, (+) and (−)
     (3) V supply pin(s) to common, reverse biased
  c. Determine (by circuit investigation, progressive increase of the ESD voltages, etching and electron beam microscope investigation) the critical paths and latent failures pattern.

The parts that have been submitted to ESD characterization should not be considered deliverable items. Therefore, ESD *is NOT a burn-in kind of test.*

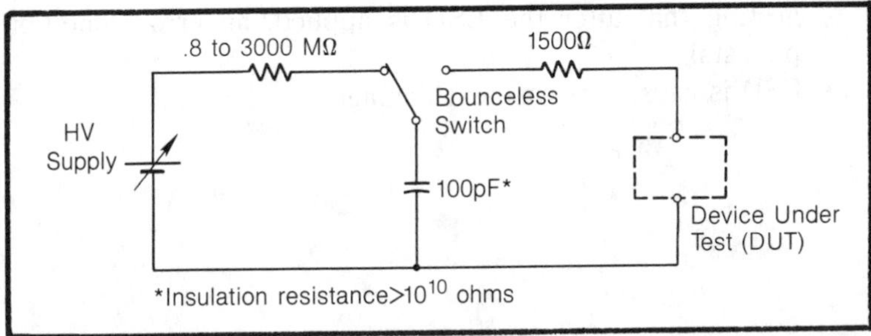

Figure 3.1—Test circuit for MIL-STD-883 and 1686

A failure is declared if, after the test:

- One electrical parameter is out of the normal device specification or,
- One electrical parameter has changed by more than 10%, even though it is still within specification limits (this indicates that the device has been permanently affected and that a potentially latent failure mode exists)

3.2 ESD Specifications for Equipment Susceptibility—IEC-801-2

Among the many standards and proposals which exist in the industry, the one which, for the moment, has become an international reference is IEC Standard 801-2, released in January, 1985. It calls for different test levels, according to the class of environment:

RH	Floor Material	Test Voltage
⩾35%	Antistatic	2kV
⩾10%	Antistatic	4kV
⩾50%	Synthetic	8kV
⩾10%	Synthetic	15kV

Recognizing the wide variety of equipments to be tested and the diversity of the operations they perform, the IEC Standard does not establish specific criteria for compliance, but recommends categorizing the severity of ESD effects as follows:

- ESD is causing a random, but repeatable event
- ESD is causing a consistent and permanent malfunction (suggesting that after the ESD is applied, an error condition persists)
- ESD is causing permanent damage.

Chapter 4

ESD Diagnosis and Testing

Testing a product for ESD vulnerability is one of the most important, versatile and easy to perform of all the electromagnetic interference (EMI) tests. Due to its wide bandwidth (⩾300 MHz), and to the strong field created locally, the test can reveal all at once many weak spots of an equipment:

- Poor PCB layout (large circuit loops, insufficient power decoupling, poor grounding)
- Insufficient immunity of I/O ports
- Missing or improperly bonded shields (long pigtails, etc.)
- Improperly mounted filters
- Missing or insufficient bonding of panels, covers, internal shields

These defects could have taken a much longer time to detect by classical methods, such as radiated susceptibility testing which requires a shielded/anechoic room, a set of transmitting antennas, and a powerful amplifier.

However, as simple as it may look, the foundation of sound ESD testing is, of course, an accurate, repeatable test set-up. Not all ESD simulators were created equal, and the test arrangements can also introduce some severe discrepancies.

4.1 ESD Simulators: How They Work

Several ESD simulators (which the EMC community quickly nicknamed "guns" or "zappers") have been marketed since 1970. The earliest ones were quite rough, but the most recent ones are generally more thoroughly designed. The improvements have aimed at the reproducibility of the rise time, the accuracy of HV setting, etc.

All of these simulators are based on the simplified models of Sec. 1.5, where the RC network is packaged in a convenient, usually hand-held, unit. A capacitor of given value is charged by a high voltage dc supply, then discharged on the Equipment Under Test (EUT) through a determined resistance. One of the high voltage electrodes, generally shaped like a finger, serves as the injection probe. The other electrode is connected to the reference ground or any desired return path.

To perform ideally, an ESD simulator should have the following features:

- variable ESD voltage, easy to set with an accurate read-out (10% accuracy is an absolute minimum, with 5% being recommended). The setting can be continuous or in 100-300 volt increments.
- an option to select a positive or negative discharge
- the ability to deliver a slope ⩾3 Amp/nanosec in a non-reactive load
- the option to select several (at least two) discharging resistors; for instance, 1000 ohms for personnel and 10 to 50 ohms for furniture.
- the ability to generate a discharge with or without an arc (in the latter, the high voltage is applied only once the probe tip is in electrical contact with the EUT)
- a shot counter so that the operator can apply a known number of discharges on each part of the EUT
- the option of selecting a simple shot or a slow repetition rate (for example, 1 to 5 discharges/sec)
- carefully selected components in the high-voltage section. All resistors, capacitors, relays, and switches must be of the high-voltage/high frequency band: no arcing, no creepage, minimum parasitic inductance, and no bouncing
- a convenient, idiot-proof design of the ground return conductor
- a filtered ac input to avoid pollution of the ac mains and ESD

problems coming in through the back door, thus yielding misleading test findings
- ease of calibration checks
- good portability and sturdy construction for intensive field usage.

Of the existing simulators available commercially, to the author's knowledge no one simulator meets all these criteria, although a few models meet almost all of them. Many differences exist between the existing simulators concerning the maximum voltage available, the making of the RC discharge network, the value of R and C, etc. Each manufacturer advocates his choices using arguments where the technical bases are not always clear. One must admit that, although the basic schematic is simple, the making of an ESD simulator is not. The designer must face all the difficulties of a high voltage transient generator, compounded by the challenge of extremely short rise times.

Also, the basic diagram bears an inherent weakness: It is presumptuous to assume that an assembly of discrete capacitors, semi-conductors, and wires will act in the same manner as the capacitance of a human body, which is distributed over a wide surface. So far only one simulator has faithfully reconstructed the body (or furniture) spread capacitance. This simulator, the IBM Type 5800 ESD transient generator, was extremely efficient but rather bulky to install and required some dexterity to use.

4.1.1 Arc or Direct Contact?

Another dilemma, not close to being resolved, is the following:
- should the ESD simulator replicate the conditions of an actual human (or furniture) event; i.e., by arcing? or
- should the ESD simulator sacrifice the arc conditions and produce, without an air gap, a calibrated pulse waveform, injected by direct contact?

The two theories have their pros and cons. The patrons of the "with-an-arc" ESD have several positive points: Arc ESD does include, naturally, all real-life arc parameters (air gap disruption, strong localized electric-field near the tip, arc resistance in the path, etc.). Very importantly, it also contains the "precursor" phenomenon, that pre-arcing occurs below 6-8 kV (see Sec. 1.5.4). And zapping with an arc can be performed rather easily on any virtually accessible point of the EUT, even recesed areas like the spaces between keys, etc.

The supporters of direct injection ESD also have many strong arguments, some of which point out the weaknesses of "arcing ESD." For instance, the arc introduces an uncertainty because the spark will not reproduce itself exactly from one discharge to another; hence, the test conclusions can be inaccurate. By eliminating the arc, the direct contact ESD guarantees a clean, "sanitized," easily reproducible waveform.

Supporters of arcing will argue that unless the discharge area is a clean metal zone, discharging on a painted or coated area *does* include an arc. Though this arc is not seen, it penetrates the paint and, therefore, still creates the same uncertainty that the direct contact was supposed to eliminate. Thus, true direct contact can only be achieved through firm electrical contact on protruding screws, keys, etc. For other areas, the operator must scratch or pierce the paint, *then* trigger the discharge. This complicates the test routine. The counterargument to this is that the so-called "easier" set-up of the arc ESD actually masks a real difficulty—regardless of the actual voltage setting on the HV dial, the arc will occur at the breakdown voltage of the air gap. In the case of repeated discharges, it is the spacing of the probe tip to the discharge surface which will trigger the arc and will, therefore, set the starting voltage. The only way to recreate actual arc ESD is to either:

- slowly approach the EUT with the probe until an arc occurs, then move back and start all over again, or
- given a specified ESD voltage, set a gap slightly larger than the breakdown distance (using 13 kV/cm as a rule of thumb) and reduce the gap gradually until an arc occurs. Then retain this setting for the given test run, or
- constantly monitor the *probe* voltage instead of merely the HV supply setting. This is what some simulators do (KeyTek, Electro-Metrics/Experimental Physics).

Another argument in favor of direct contact ESD is that it eliminates the operator's body effect. In arc ESD, the added stray capacitance of the operator may influence the test waveform, especially if he stays close to the probe tip. Finally, arc ESD requires a controlled relative humidity during the test (air ionization can be influenced by the RH), while direct contact ESD is rather insensitive to RH.

A whole chapter could be filled with all these arguments. In an attempt to clear the waters, Table 4.1 lists seven pros and cons and gives to each item a (+) or (−) score, weighted by its relative im-

Table 4.1—Arc discharge vs. contact discharge pros and cons

Each topic is either a bonus or a penalty, but is never counted on both sides. A 1 (less important) to 3 (most important) weighting factor has been used.

	Arc ESD	Direct Contact ESD
Replicates real-life arc condition including pre-arcing	+3	
Guarantee of a well-reproducible waveform		+3
Cannot be performed on hard-to-reach recessed areas		−3
Requires scratching the paint		−2
Requires a carefully set gap dimension	−1	
Insensitive to operator's body proximity and other human factors		+2
Insensitive to RH		+1
TOTALS:	+2	+1

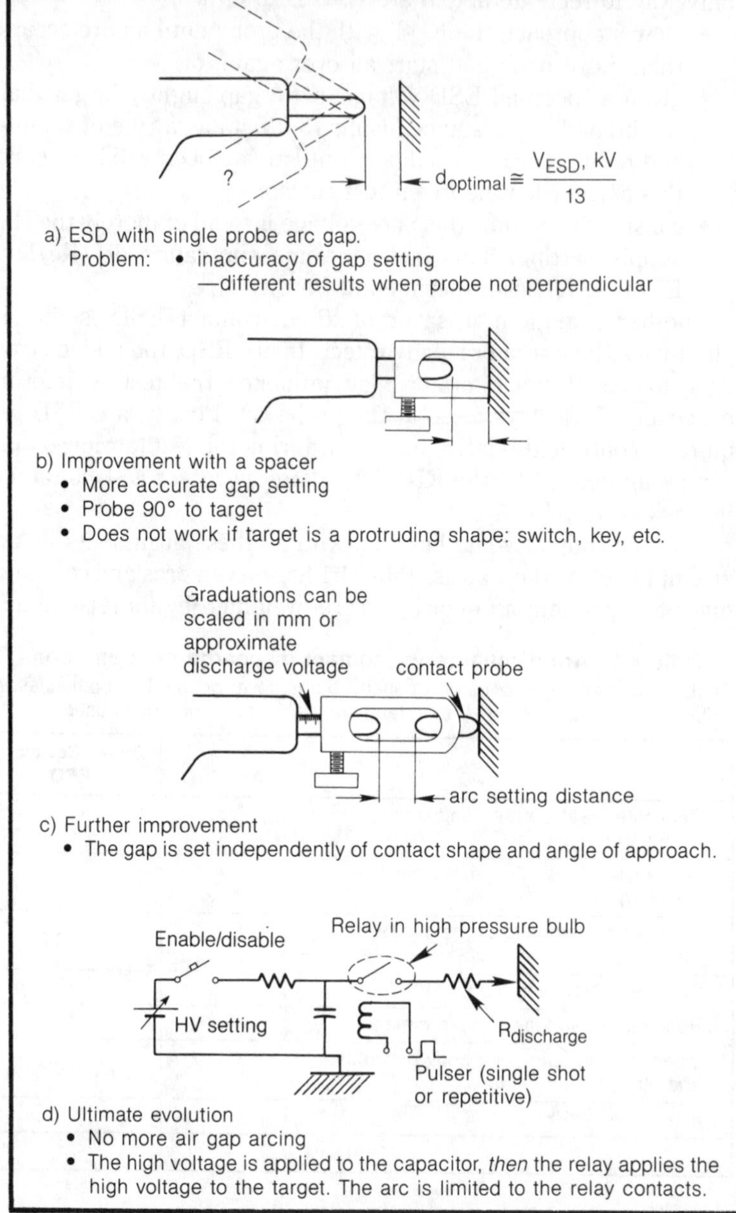

?
$d_{optimal} \cong \frac{V_{ESD}, kV}{13}$
a) ESD with single probe arc gap.
Problem: • inaccuracy of gap setting
—different results when probe not perpendicular
b) Improvement with a spacer
• More accurate gap setting
• Probe 90° to target
• Does not work if target is a protruding shape: switch, key, etc.
Graduations can be scaled in mm or approximate discharge voltage
contact probe
arc setting distance
c) Further improvement
• The gap is set independently of contact shape and angle of approach.
Enable/disable
Relay in high pressure bulb
HV setting
$R_{discharge}$
Pulser (single shot or repetitive)
d) Ultimate evolution
• No more air gap arcing
• The high voltage is applied to the capacitor, then the relay applies the high voltage to the target. The arc is limited to the relay contacts.

Figure 4.1—Evolution of simulator probe concepts

portance. Although the score unavoidably contains some subjective biasing of the author, the two methods come very close. Indeed, the issue is one of a test philosophy: Should a test try to replicate by all means the conditions of the actual event at the expense of losing some reproducibility? Or should a test be aimed at generating a calibrated, repeatable stimulus that will stress the EUT "as if" it were the actual event, even though the electrical mechanisms are not all there? The personal opinion of the author is that the latter philosophy should prevail.

The next section shows how some typical ESD simulators available in 1985 have handled these conceptual problems.

4.1.2 Simulators Based on IEC-801/2 Recommendation

Since it is likely that IEC-801/2 (previously IEC.65.ESD) enjoys international acceptance, we will briefly describe the simulator stipulated by this document. Fig. 4.2 shows its main features. The simulator uses the simplified equivalent circuit of personnel ESD (see Sec. 1.5), but with a discharge resistor of 150 Ω instead of the 1 kΩ or more typical of human body models. The storage capacitor is charged through a 100 MΩ limiting resistor which corresponds to a charging time constant RC of 15 msec so that after about 0.1 sec (more than 6×RC), the capacitor is considered fully charged.

The basic schematic simply requires a switch on the source side and does not require a transfer switch. To trigger the arc discharge required by the document, the discharge surface is slowly approached by the probe until arcing occurs.

4.1.3 Critique of the IEC-801/2 Type of Simulator and Test Practice

The IEC type of ESD simulator, along with its companion test procedure, is gradually gaining the advantage of being an internationally accepted test circuit. This international specification, with its broadly accepted values and test set-up, provides a common base-

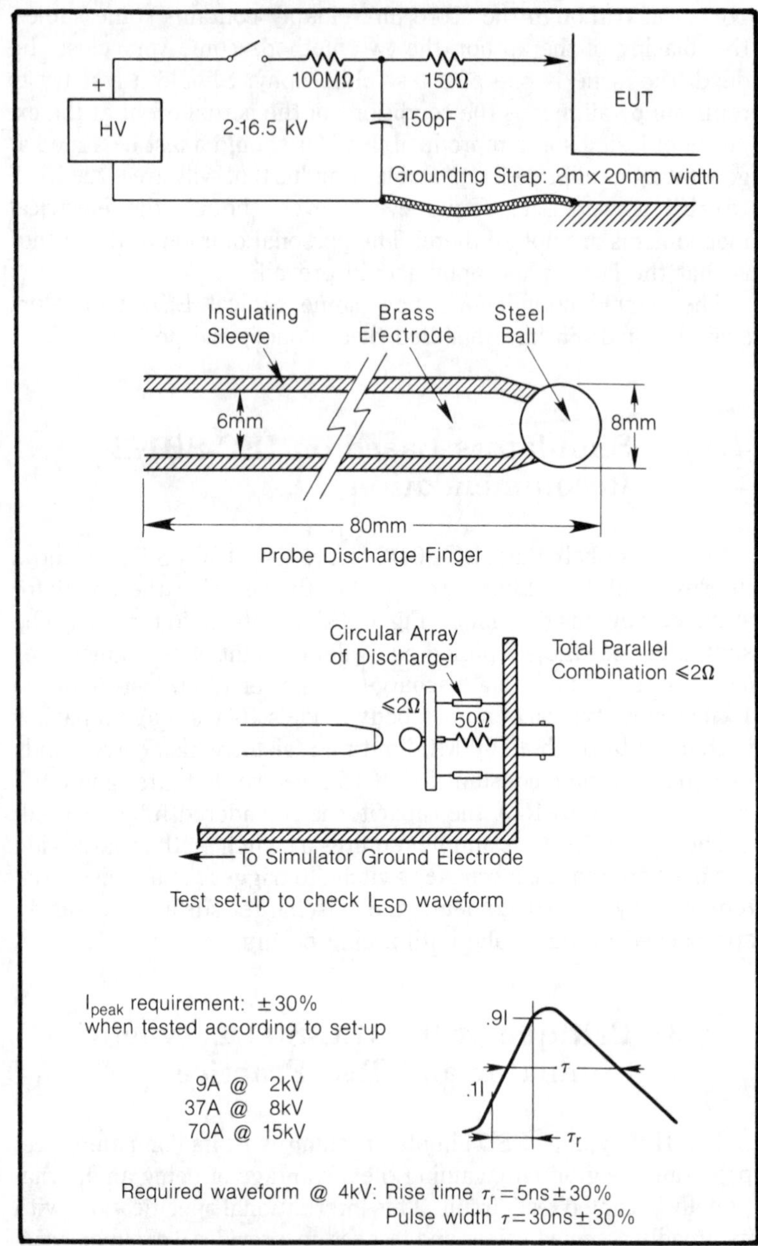

Figure 4.2—IEC-801/2 simulator

line for assessing the ESD immunity of many products. (Not always remembered, though, is the fact that the 1985 edition of IEC 801/2 does not claim to address all sorts of electronic equipment, but simply "industrial process control equipment.")

The IEC-specified simulator and its installation do have a few shortcomings of which the user should be aware. Armed with the discussion about "arc vs. no-arc" (Sec 4.1.1), the reader will probably identify some of the latent weaknesses of the IEC-type simulator (at least in its 1985 definition). In addition to the problems pertinent to the choice of an air gap discharge, a few others are peculiar to the IEC document itself.

- The rise time and peak current specified by IEC indulge the simulator's inaccuracy too much. Par. 6.2 specifies that the rise time at 4 kV be 5 nsec ±30% and that the corresponding current be 18 A ±30%. Therefore, an ESD gun "A" delivering 23.4 A with 3.5 nsec rise and an ESD gun "B" delivering 12.6 A with 6.5 nsec rise are both said to comply with IEC-801. However, their dI/dt being 6.68 A/ns and 1.94 A/ns, respectively, the faster one will create induced effects 3.4 times more severe than the slower one!
- The "precursor" effect, one of the features of the arc ESD, has been circumvented in a rather cavalier manner. Par. 2 says that ". . . rise time and current of the pre-discharge are not measured nor evaluated by the testing procedure."
- The choice of a unique discharge resistance of 150 ohms is a trade-off. Being midway between a severe human ESD (the lowest range of body resistance is about 1 kilohm) and a furniture ESD (very low impedance), it has the advantage of simplicity. But it lacks the important aspect of furniture ESD; i.e., the oscillatory waveform of the under-damped RLC network (see the discussion in Sec. 1.5). It is not clear whether this shortcut is supported by thorough studies or is an arbitrary choice made for the sake of simplification. In any case, the reasons for it are not documented in the IEC recommendation, and the user is left in doubt.
- The pulse is positive only.
- The requirement for the ESD current return path is rather sloppy. (Discussion of the ground wire placement is deferred to Sec. 4.2.) But the length of this ground wire bears some serious drawbacks which impede the whole validity of the test.

The detailed calibration procedure of the gun in Par. 6.2 of the IEC standard specifies that when discharged on a quasi-short circuit made by a 2-ohm coaxial resistor and a metal bracket, the discharge characteristics should be centered on:

$$18\ \text{A at } 4000\ \text{V with } \tau_r{=}5\ \text{nsec}$$

Therefore, the total impedance of the test circuit is:

$$Z_t = Z_s + Z_d = \frac{4000\ \text{V}}{18\ \text{A}} = 222\ \Omega$$

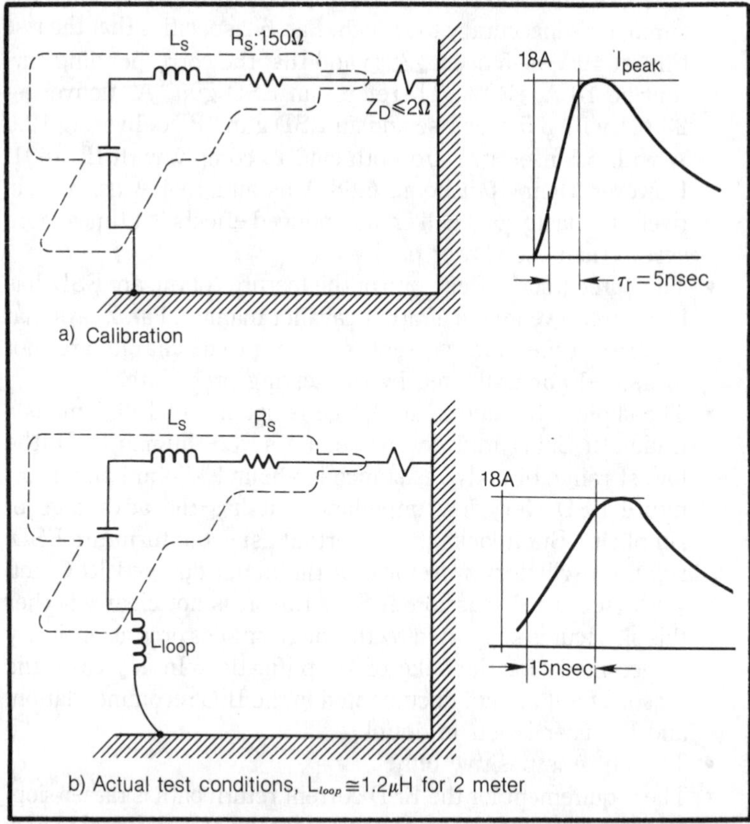

Figure 4.3—Problems with the parasitic impedance of the simulator ground strap

with

Z_t = total impedance
Z_s = internal impedance of the simulator
Z_d = impedance of the discharge calibration circuit

Since Z_d is made of a 2-ohms non-inductive resistance and a short, wide, metal bracket with negligible impedance, we can approximate:

$$Z_s = 220 \text{ ohms}$$

where Z_s consists of the gun resistance and series inductance L_s. Thus,

$$Z_s = 220 \text{ ohms} = \sqrt{X_s^2 + R_s^2}$$

with R_s=150 ohms, the specified discharge resistance. Solving for X_s gives:

$$X_s = 161 \text{ ohms, with } X_s = \frac{L_s}{dt}$$

$$L_s = 161 \text{ ohms} \times (5 \times 10^{-9} \text{ sec}) = 0.8\ \mu\text{Henry}$$

Once the gun has been calibrated this way, the actual ESD testing is specified with a 2 m long ground return strap located 10 cm from the EUT. There is no way on earth that such a loop can make less than 0.6 μH/meter; i.e., a minimum of 1.2 μH for 2 meters. Therefore, the reactance added by the IEC prescribed ground strap is:

$$\frac{L}{dt} = \frac{1.2 \times 10^{-6}}{5 \times 10^{-9}} = 240\ \Omega$$

The total loop impedance during an actual test will be:

$$Z_{tot} = \sqrt{[R_s^2 + X_s^2] + X^2_{strap}} = \sqrt{(220)^2 + (240)^2} = 325\ \Omega$$

Then the (apparent) discharge parameters for 4kV:

$$I_{peak} = \frac{4000}{325} = 12.2 \text{ A (instead of 18)}$$

Since with such an inductance, the rise time will no longer be 5 nsec, we can evaluate it:

$$\text{Rising Time Constant: } \frac{L}{R} = \frac{0.8\ \mu\text{H} + 1.2\ \mu\text{H}}{150\ \Omega} = 13 \text{ ns}$$

$$\text{Current Rise Time, } \tau_r \text{ (10\%-90\%)} \cong 1.2 \times \text{Time Constant}$$
$$= 15.6 \text{ ns (instead of 5 nsec)}$$

This now gives a total circuit impedance R+jx of 220 Ω. The specified 18A are there, but with a slope of 1 A/nsec, quite different from the 3 to 10 A/nsec or more which characterizes ESD! A test made like this would underestimate actual equipment vulnerability.

These defects are correctable. Until the IEC publication is revised, the user who is aware of these issues can overcome them. It is recommended at least that:

- when calibrating the simulator, one should not be so lenient as ±30% accuracy on I and τ_r and aim for ±10%.
- when testing, the ground strap should be made as short as possible; for instance, a 20 mm wide, .80 meter long strap would upgrade the rise time to 10 nsec.

Figures 4.4 and 4.5 show two modern ESD simulators based on IEC-801/2 made by two well-known manufacturers of high-voltage generators. The simulator in Fig. 4.4 is made by Schaffner. It is extremely handy, easy to hand-carry on an airplane for short-notice field trips and is rated up to 16 kV (20 kV for model NSG 431). Its discharge probe is strictly based on the IEC prescription. Its installation and usage are remarkably easy.

Figure 4.5 shows the model 2000 simulator manufactured by KeyTek. It is a more complex instrument; therefore, it is heavier and requires some experience to be used properly. On the other hand, it is a very complete tool, with a lot of built-in features which make the test waveform accurate and reproducible. It also has a set of several discharge heads to enhance either the E or H-fields. Finally, a direct contact discharge is also feasible with a special head including a high voltage relay.

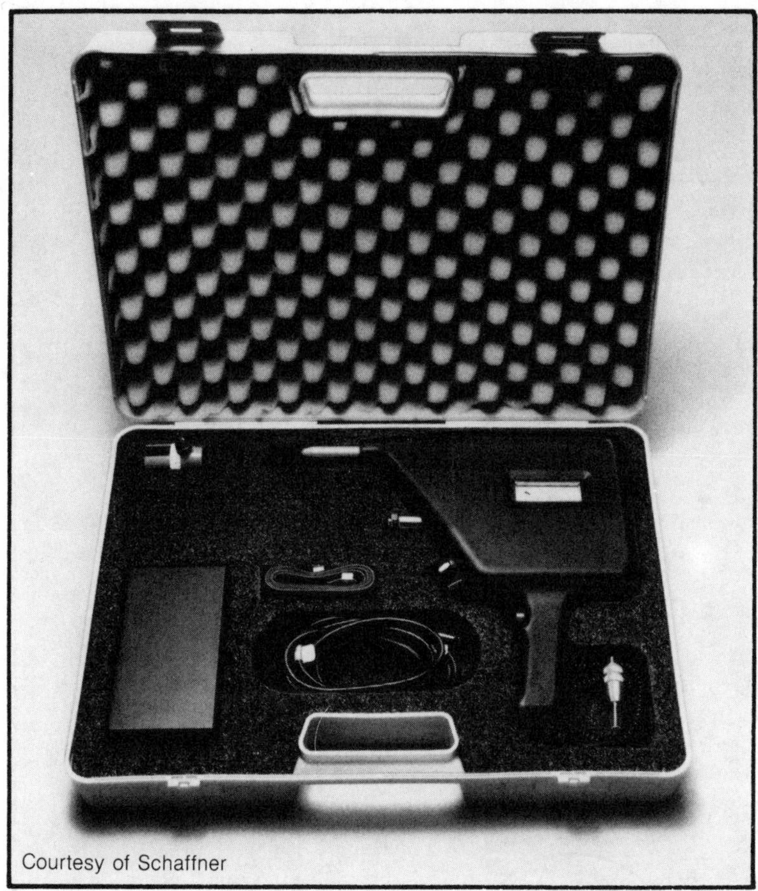

Courtesy of Schaffner

Figure 4.4—The Schaffner gun

4.1.4 Other Types of ESD Simulators

There are no fundamental differences between the type of gun described by IEC publication 801/2, and all the other types of ESD simulators available on the market today for equipment testing. Differences lay principally in the different RC networks required by specifications other than the IEC, the sophistication of the features, the options offered (see the list in Sec. 4.1) and whether the simulator can create a direct contact (no air gap) discharge. In a

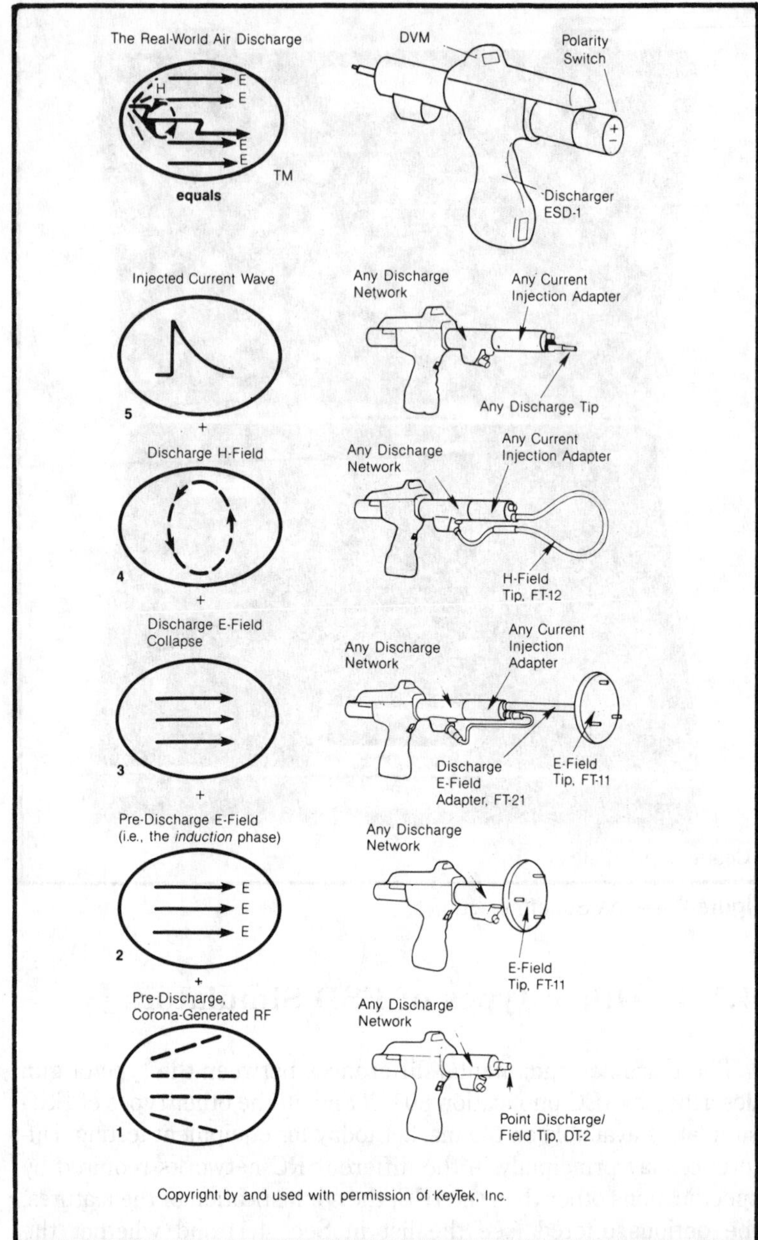

Figure 4.5—The KeyTek gun, simulating the basic aspects of ESD

few cases, such as IBM model #5800, manufacturers of equipment have developed ESD simulators for their own use which differ significantly in concept from the IEC model. But these devices are generally not sold commercially.

However, there are ESD testers which are used quite differently. These are the ESD simulators used to determine *component vulnerability*, per MIL-STD-883 or 263, for instance. The purpose of such equipment is to inflict a given ESD level to every possible pin combination of an integrated circuit (or, eventually, any ESD sensitive component). These tools no longer look like hand-held guns, but rather they resemble work-stations with programmable module sockets as in Fig. 4.6. The selected pin(s) and the test sequence can be fully or semi-automated. The test results can be visualized on a separate curve-tracer or, sometimes, on a CRT which displays the current vs. voltage (I,V) curves of the stressed device.

On the more sophisticated models, like the one seen in Fig. 4.7, the user can program a "halt" after any given type of failure or obtain an overall print-out in the form of a bar-chart, etc. (see Fig. 4.8).

All these sophisticated features may seem like luxuries; however, consider a manufacturer who has to regularly monitor the ESD sensitivity of his products. Consider, for example, an integrated circuit with 24 pins. Each pin must be tested, for example, against all the other pins grounded together. That makes 24 tests with both polarities. And since what is looked at is a quality figure, a statistical distribution of the failure levels is necessary. Therefore, the voltage

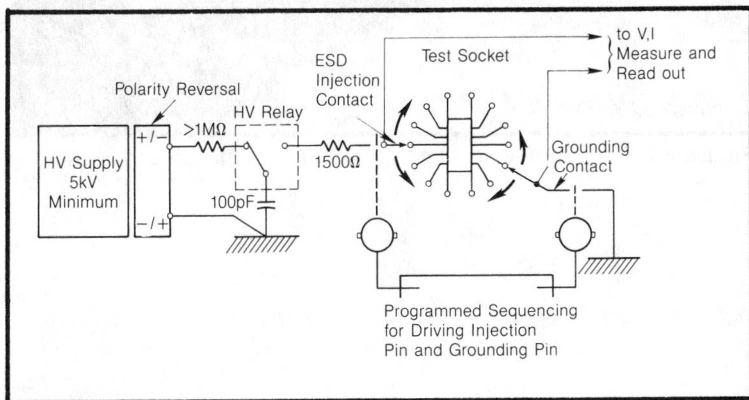

Figure 4.6—ESD simulator for automatic component ESD testing

will be gradually increased from 500 V to 4000 V in 250 V steps. Assume also that the specification requires each pulse to be repeated five times. The total number of measurements, depending on the spread of the devices' characteristics, will vary between:

$$24 \times 2 \times 5 = 240$$

for those devices failing after the 1st voltage step, and

$$24 \times 2 \times 5 \times \frac{4000\ \text{V} - 500\ \text{V}}{250\ \text{V}} = 3360$$

for those devices which resist up to 4000 V! It is, of course, unrealistic to have these tests done manually.

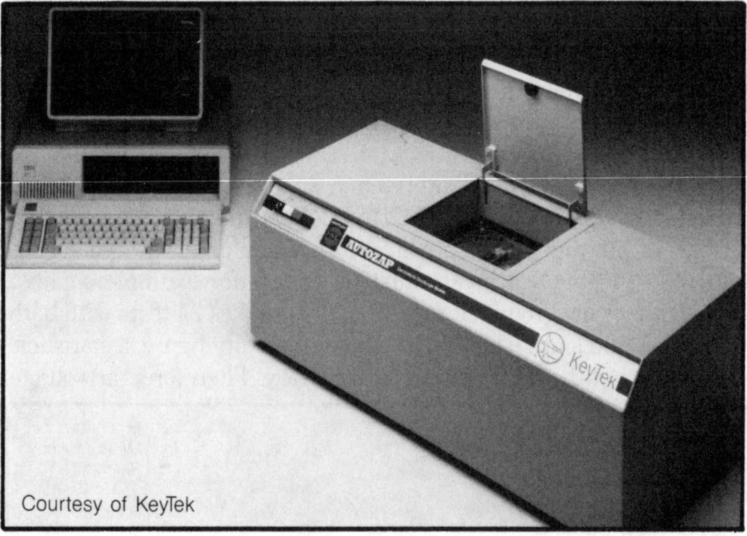

Figure 4.7—The Hartley simulator (distributed in the USA by KeyTek)

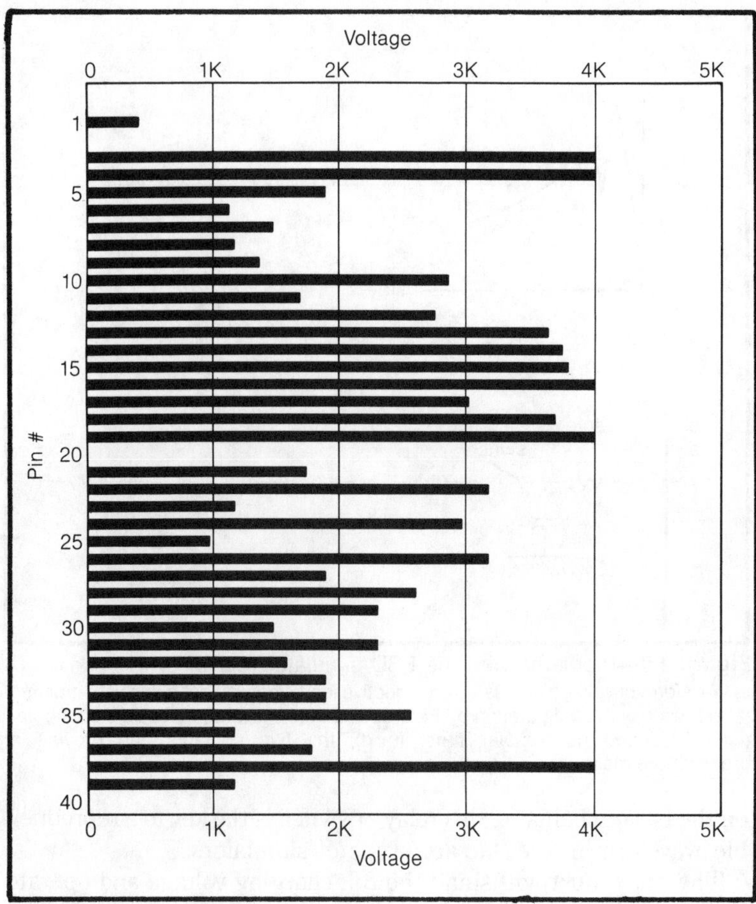

Figure 4.8—Example of the bar-chart print-out delivered by the "Autozap" tester = ESD damage thresholds for an 8085 integrated circuit

4.1.5 Special Relays Required for ESD Simulators

For both the component ESD testers of Sec. 4.1.4 and the most sophisticated ESD guns implementing the direct contact (no-arc) philosophy, a special relay is needed to commute the capacitor end from the charging to the discharging mode. Although it looks trivial

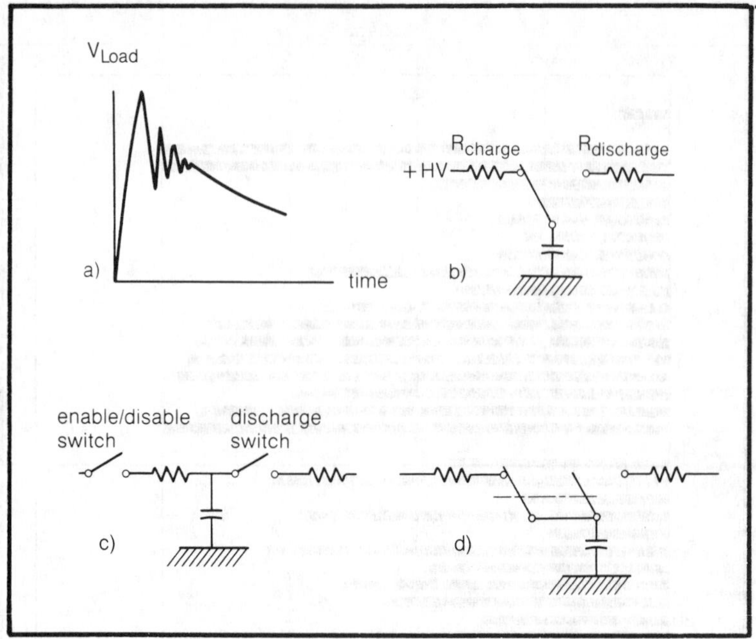

Figure 4.9—Problems with the ESD simulator relays
a) Waveform of voltage across a non-inductive load, showing bouncing with ordinary relays, used with configuration b). Figures c) and d) show other possible configurations to speed up transfer time. In d), the two contacts can provide a break-before-make.

on the basic schematic, this relay or switch is the key to a reproducible waveform in the "no-arc" type of simulators.

The relay must withstand the full charging voltage and operate the transfer without bouncing. Also, although direct contact was supposed to eliminate the arc problem, nothing is free and there is still an arc, but now it is in the relay. Until the time comes when solid state switches with 1 nsec transition times will handle 20 kV, the only way to lessen the arc problem is to use high-voltage relays with minimum arcing and bouncing. For instance, mercury relays like those made by Magnecraft/USA can be used successfully up to 10,000 volts or more. Other manufacturers use relays where the contacts are in a sealed bulb with a high pressure gas, so that a 1 mm gap will withstand 10 kV, for instance (see Paschen curves, Appendix C). Branberg (Ref. 22) reports good results from a high-voltage relay with pressurized SF_6 gas purchased under the Hewlett-Packard reference #0490-0655.

In his very complete study (Ref. 26), S. Vrachnas found that:

- *vacuum relays* produce unacceptable bouncing
- *reed switches* may have high voltage withstanding but can produce waveform alteration due to their parasitic capacitance
- *mercury relays* give good results. However, 10-15 kV is their upper limit by today's standards.

Finally, he concludes by showing that, rather than eliminating the arc, a well controlled triggered spark gap (EG&G type GP92) can also yield an acceptable waveform, but that the range of possible voltages is limited to the specific type of gap used.

Other authors, like Mazdy (Ref. 9) as well as Shaw and Enoch (Ref. 27) have reported the same problems associated with the relays. The ideal switch is certainly technically feasible but the market of ESD simulators is probably too small to justify a special study by the relay industry. So the designer of an ESD tester is left only with what is available.

4.2 ESD Test Set-Up

A basic test set-up for ESD testing is shown in Fig. 4.10. The equipment under test (EUT) is normally installed above a reference ground plane at a height dictated by its wheels, feet, casters, etc. The return conductor (preferably a strap) of the simulator is grounded to this reference plane, and the ESD is applied to all points of the EUT likely to be touched by personnel and furniture.

The following sections describe the details of the test procedure. When there are disparities between what is recommended here and what some of the principal specifications require, this will be mentioned.

4.2.1 Ground Reference

A reference ground plane is of *prime importance* and must be installed every time to stabilize the EUT-to-ground capacitance, thereby allowing repeatable testing. Without this plane, the return path of discharge would be totally uncontrolled, shared between the undefined earth plane and the room ground wire; thus, the rise time and the spread of the return currents would differ, and the test results would not be reproducible from one installation to another. The ground plane can be made of a copper or aluminum plate or foil which must extend around the EUT 10 cm or enough to encompass both the EUT (which can be a system of several units) and the simulator.

To prevent the ground plane from rising to an undefined potential after many successive discharges, and possibly creating testing inaccuracy or even shock hazard, the ground plane must be connected to the local earth wire via a resistor of about 1 kilohm. This resistor is used to avoid uncontrolled return paths in the loop formed by the EUT, its power cord, the earthing terminal, and the ground plane. This loop, being uncontrolled, could create repeatability problems.

The EUT-to-ground plane installation can be more complex than the simple sketch of Fig. 4.10. The EUT can be an upright, floor-standing machine or a table top equipment. It may have a metallic housing or a non-conductive one. The following recommendations are made for the various options (see Fig. 4.11):

A. For equipment with metallic housing

A.1. If floor-standing, the equipment should stand over the ground floor as shown in the basic configuration of Fig. 4.10.

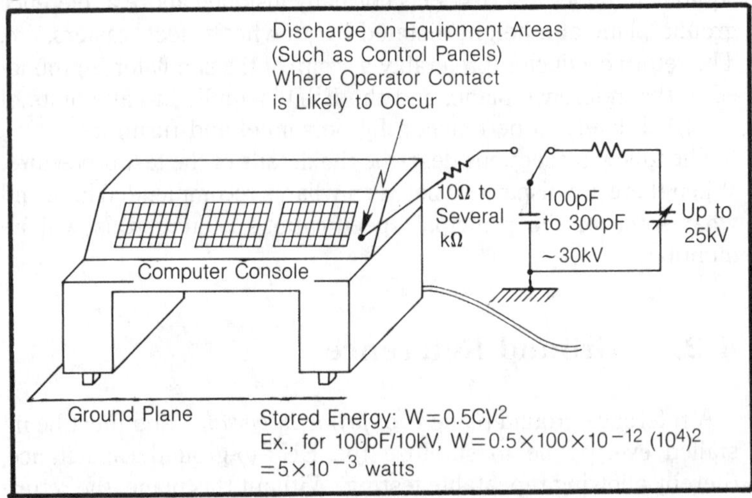

Figure 4.10—Basic test set-up for equipment testing

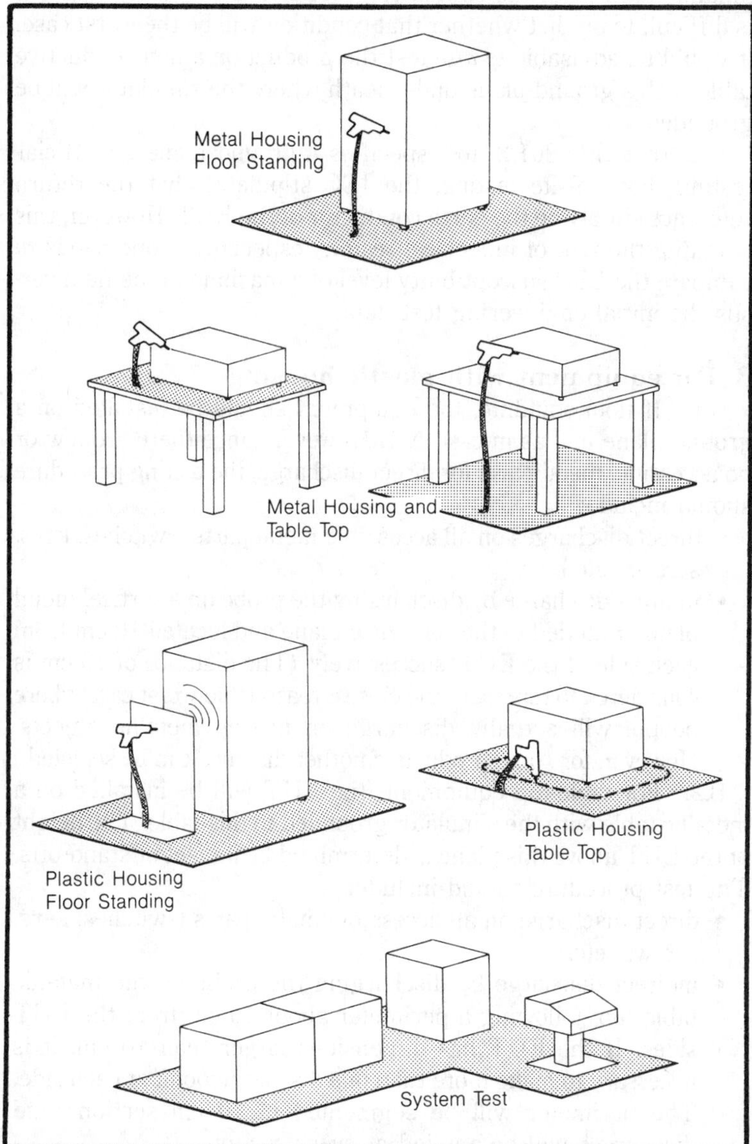

Figure 4.11—ESD test set-ups for various types of machines

A.2. If it is a table-top equipment, the EUT should stand on a metallic table with the simulator grounded to this table. Since it is difficult to predict whether that condition will be the worst case, it would be advisable to *also* test the product on a non-conductive table with a ground-plane underneath where the simulator will be grounded.

Note that IEC 801/2, too, specifies a ground plane for official testing. For on-site testing, the IEC stipulates that the return reference simply be the earth conductor of the EUT. However, this is taking the risk of indecisive results, especially if one needs to compare the ESD susceptibility level of a machine in the field versus the initial engineering test data.

B. For equipment with plastic housing

B.1. If floor-standing, the equipment should be installed on a ground plane just as in case A.1. However, since there are few or no accessible metal parts for direct discharge, the testing procedure should include:

- direct discharges on all accessible metal parts (switches, keys, screws, etc.)
- indirect discharge by discharging the probe on a vertical metal plate grounded to the reference plane and located 10 cm from each side of the EUT, successively. (The distance of 10 cm is considered to represent the closest reasonable worst case where people will actually discharge on nearby metallic objects. However, for some products, another distance can be selected.)

B.2. For table-top equipment, the EUT will be installed on a metallic table with the simulator grounded to this table. The height of the EUT above this plane is determined by its feet or stand-offs. The test procedure should include:

- direct discharge on all accessible metal parts (switches, keys, screws, etc.)
- indirect discharge by discharging the probe on the metallic table top, following a perimeter about 10 cm from the EUT sides. If the EUT has dimensions larger than 10 cm, it is necessary to make more than one discharge point at each side. The perimeter will be segmented in 10 cm sections and discharges will be applied at every section.

Although not as critical as for A.2, it would be advisable to *also* test the product on a non-conductive table, with a ground plane underneath where the simulator would be grounded. In this case, the direct discharge would be applied to the accessible metal points and the indirect discharge to a vertical grounded plate, as for B.1.

This requirement of testing metallic and plastic table-top equipment with both conductive and non-conductive tables seems extravagant, but experience has shown that depending on the orientation of PCBs inside the unit, the points of entry and internal routing of the cables, one of these configurations may be more critical than the other. To save test time, one can always run an exploratory test with both configurations and find which one gives the weakest ESD fail levels. Then retain that set-up for the rest of the test.

To be consistent with the ground plane requirements, the metallic table should be of aluminum. If it is not, an aluminum or copper foil should be laid over the table top. The earthing requirements for the metallic table are the same as those for the ground plane.

4.2.2 Grounding of the Simulator and the EUT

The EUT should be earthed through its normal earthing conductor to the earth terminal of the ac outlet. If the EUT has no earth conductor (class II equipment) or is battery operated, it will then be un-earthed. Under no circumstances should the EUT be directly grounded to the reference plane, unless this is the way it would normally be installed in the field.

The simulator return for ESD pulse must be a large strap, preferably no longer than 1 meter (see the discussion in Sec. 4.1.3 about the crucial effect of this length on rise time), bonded to the reference plane via a "C" clamp or a similar type of clamp. *The simulator should not be grounded to the EUT frame, except for some diagnostic testing.*

Note: This deviates from the IEC test method, which shows the simulator being grounded to the EUT in certain cases. This is questionable since, in real life, charged people will have their feet on the ground, and not on the EUT cabinet. In addition, grounding the simulator to the EUT frame will decrease, or even nullify, the electric field change between the lower part of the EUT (especially the bottom plane) and the ground plane. This field change is important in the possible coupling to I/O cables near their entry points.

4.2.3 External Cables and System Configuration

Very often, the ESD test may have to replicate a system configuration. There can be several reasons for this. For instance, the EUT may be one unit of a multiple box system, in which case the vulnerability of the whole system must be evaluated by testing one box after the other. Alternatively, the EUT may be designed as a stand-alone that is "attachable" to several types of peripherals or ancillary equipments. In this case, the peripheral devices must be connected to the EUT, even if they are not themselves being tested. However, it is necessary that the units not being tested have an ESD level consistent with the test objective.

For a system configuration, the whole system must be installed over the artificial ground plane. If this is not practical, one ground plane per box can be used, provided that these planes are interconnected by wide straps.

All the external cables which can be connected to the EUT in its maximum configuration should be in place and should be laid out in a typical installation arrangement. To avoid too much variation in results due to different heights above ground, they should be laid out at a constant, repeatable height above the ground plane. Unless other heights are dictated by specific applications, a height of 10 cm, using wooden or plastic spacers, seems a good average (see IEC-801).

4.3 ESD Test Routine and Discharge Procedures

The following describes the recommended ESD test routine.

1. First, make a zoning by dividing each side of the equipment into approximately 0.1m²(30 cm×30 cm) areas. Mark/code each area. Include signal cables and power cord entry areas.

2. Next, determine a clear, indisputable *malfunction* status which can be recognized *without the need of an external oscilloscope, data-logger, etc.,* such as hard-error, wrong read-out, inadvertent reset, alarm, power-down.

This point is very important. No external ancillary equipment should be used to diagnose a fault condition, because the very presence of additional probes and cables, and the monitoring device itself can cause the EUT to fail at lower levels and give wrong test results. Very possibly, the scope or data-logger can itself be disturbed by the ESD test and give misleading information.

Therefore, if the EUT is a programmable device, it may be necessary to develop a software routine which:

- exercises continuously all EUT operations in closed-loops without requiring operator intervention
- indicates clearly by a print-out, CRT message, alarm, buzzer, indicator light, etc., that a fail condition exists

3. After the malfunction status has been determined, set the ESD level at about 5 kV (for a personnel type discharge) or some other determined value, depending on whether the routine is an investigation or a QC test.

4. Zap each of the coded areas marked in the first step. If this area includes switches, keys, indicators, connectors, screws, rivets, etc., apply the discharge on those points. If there are no such elements, apply the discharge on the seams, slots, and any protruding shape, angle or surface discontinuity existing in the area. Otherwise, simply apply the discharge in the middle of the coded area.

If the housing is plastic, the test should use an indirect ESD. However, a direct discharge should still be tried on screws, rivets, decorative trims, etc.

5. Repeat the discharge until the prescribed number of pulses have

been applied without an error. If no minimum number of pulses is prescribed, use 50 as a default. However, the minimum number of pulses to apply to guarantee the test depends on the complexity of the EUT operations (see Fig. 4.12). Section 4.6 discusses in more detail the problem of the number of pulses to apply. If a dual polarity discharge is prescribed, repeat with reverse polarity.

6. Repeat step 5 for all coded areas and record which ones failed. If none failed, increase the level by 1 kV and re-run the test.

7. For each failed area, decrease the ESD level to find the threshold of GO/NO-GO, and record.

8. Starting with the weakest spot, apply EMI hardening methods described in Chapter 5 to meet 10 kV, 15 kV or other objective.

Some simulators have a selectable repetition rate. This is a useful feature because manually applying 50 discharges times "n" points in a single-shot mode would be tedious. However, the repetition rate must be set sufficiently slow to allow for software recoveries (if the EUT has such features); otherwise, the unit would lock-up in an error mode which would be misrepresentative of real ESD situations where there are never several events per second. Another phenomenon may occur which can also foul up the test if a too-fast

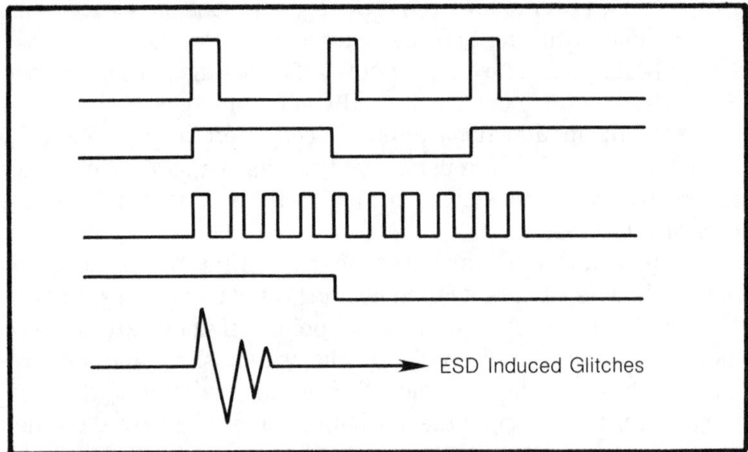

Figure 4.12—The ESD induced transient is a random event which occurs anywhere vs. the sequence of logic operation and message formats of the machine under test. A minimum number of discharges is necessary to explore the worst-case incidences of the ESD transient with certain patterns of logic transitions.

discharge rate, such as 50 pulses per second or more, is applied. Some devices have a very high input resistance; if, in addition, their 0-volt reference has been floated to avoid ground loops, the ESD bleed path can be quite intricate and the RC discharge time constant to ground can be quite significant—up to several milliseconds. This means that the first discharge may induce a noise voltage which is not in itself sufficient to upset the device but which, because it does not immediately sink to ground, leaves a residual voltage when the next pulse occurs. The residual voltage is added to the voltage from the new pulse and so on. After a certain number of pulses, this gradually reaches the upset level and creates an error or malfunction which is mistaken as a "true" ESD failure. The typical symptom in this case is an EUT which will resist the first discharges, but which will fail *regularly* after the same number of pulses, even if the voltage level is slightly reduced. Therefore, because of the two mechanisms mentioned above, it is a safe practice to check, by slowing down the repetition rate, whether the *"fail" level for a given number of pulses is correlated with the pulse rate.* For the validity of the test, what is important is the minimum number of pulses applied—not their repetition rate.

To help document the test results in an orderly manner, the test log form of Fig. 4.13 is recommended. It will allow one to keep an accurate track of the failing zones *and record the fixes that worked and those that did not.* Too often, this is neglected and people have to "re-invent the wheel" at every test. A dual level "RUN/FAIL" is also recommended to show the level at which the EUT undoubtedly meets the criteria and the level at which it fails.

4.4 No Error/No Damage Concept: The Several Layers of Severity

Like any surge-type test, testing for ESD involves threshold criteria. However, before the test is made, the designers of the EUT as well as the test personnel must *clearly* define what is to be considered a failure. Too many times, statements are made like, "our system has been tested to 8 kV ESD." Does this mean:

- that the system does not exhibit ANY malfunction up to 8 kV?

EUT: ____________________ Test Date: ____________________

Test Type { Prototype ____________________ / Release Test ____________________ / QC Test ____________________

EUT Program Used: ____________________

Simulator Type: ____________________ Checked: ____________________

Minimum No. of Pulses: ____________________

Sketch EUT Test Points

1 2 3 4 5

Test Point No.	Discharge Resistor Value	Polarity +/–	$V_{RUN/FAIL}$	Observations (Failure Mode, fixes, etc.)

Figure 4.13—Suggested ESD test log form

- that the system does exhibit malfunction starting at 8 kV? If so,
  - were these malfunctions only soft errors?
  - were they hard errors?
  - were they solid damages, requiring component replacement?

To clarify the situation, Fig. 4.14 shows a three-level ESD criterion. No voltage has been put on the scale since it will depend on the type of product and the type of market. For instance, in selecting the ESD voltage levels of Fig. 4.14 one should consider:

- the likelihood of high human activity around this product
- the type of environment (controlled or uncontrolled RH, anti-static floor carpeting)

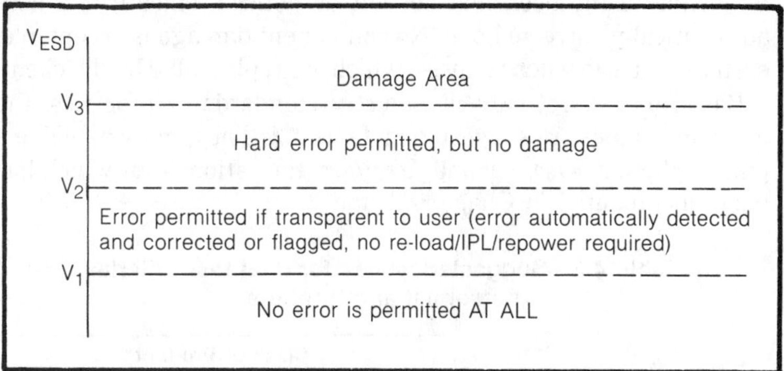

Figure 4.14—Multi-level ESD test criteria

- the sensitivity of the user to a temporary malfunction or error; i.e., how often a week or a month can a malfunction occur and be considered "tolerable" by the user? This can depend on the price range of the product, the seriousness of a temporary data loss for the customer and, finally, the general EMI immunity claimed in the functional specifications.

V_1, the lower level, is the one up to which no malfunction *at all* (recoverable or not) is tolerated. Consider, for instance, an airline reservation terminal which exhibits recoverable errors for 3 kV personal ESD. The number of ESD events exceeding 3 kV in a counter type of environment is very high (see Sec. 4.1). Even if the ESD induced errors are automatically detected and the transaction is cancelled and then reiterated, these operations take some time and this terminal will spend too large a percentage of time recovering from errors, especially during the winter/spring season.

Between V_1 and V_2, errors are permitted if they are "transparent" to the user, i.e.:

- they are self-corrected
- they do not require a user intervention (halt-restart, data re-entry, program reload, etc.) to restore normal operation.

Above V_2 and up to V_3 "hard" errors are permitted; i.e., this level is sufficiently high to have a low probability of occurrence which will not upset the user if operator intervention is necessary to restore the system operation. However, it may be required that the error be visible to the user, and not left un-noticed, since it is a non-automatically corrected one. No component damage is acceptable, even those parts (such as fuses) which are replaceable by the client.

Based on several current industry standards or practices, the following values are suggested in Table 4.2. They imply a dual test philosophy for *personnel* and *furniture*, the rationale of which has been documented in Chapters 1 and 2.

Table 4.2—Suggested values for dual test criteria: personnel *and* furniture

	Class of Product			
	A	**B**	**C_1**	**C_2**
Up to V_1 Personnel/Furniture No error or malfunction at all	NA	4kV/1kV (25A)	3kV/800V (20A)	5kV/1.3kV (33A)
Up to V_2 Personnel/Furniture Only recoverable errors	10kV/2kV (50A)	15kV/3kV (75A)	10kV/2kV (50A)	22kV/4.5kV (110A)
V_3 Personnel/Furniture Hard errors, but no solid damage	20kV/4kV (100A)	25kV/5kV (125A)	25kV/5kV (100A)	30kV/6kV (150A)

Notes:

Category A: Home Appliances and Entertainment Devices

High human activity, uncontrolled environment. Aggravating factors (e.g., dust and friction with vacuum cleaners, cloth friction with washers and dryers, etc.) but utilization is rather fault-tolerant as long as there is no solid damage.

Category B: Office Products, Home Computers and Point-of-Sales

High human activity, uncontrolled environment. Irritability factor quite high. RH can go as low as 15%, and any type of carpet may be involved. However, partial alteration of data not catastrophic because generally detectable by operator or user.

Category C_1: Large Business Computers, Computing Centers, Teleprocessing Systems Handling Critical Data (Banks, Government, etc.). C_2: Large Industrial Process Control, and Airline Reservations Systems.

V_2 level corresponds to less than .1% of ESD events during the worst season. Considering that this type of equipment has about 3 to 10 times less human activity than categories 1 and 2, this corresponds to less than one risk of hard error/month during the worst season, for the worst installation. The furniture discharge network corresponds to an equivalent impedance of 40 ohms for the first pulse of the oscillatory discharge. Category C_1 is a controlled environment, with RH>35% and anti-static carpet. Category C_2 is an uncontrolled environment.

In cases where the IEC type of simulator must be used, Table 4.3 shows recommended values for a single discharge network, although a real equivalence of both personnel *and* furniture is impossible to cover with one single circuit.

Table 4.3—Suggested values for a single test criteria, based on IEC discharge network,* 150Ω/150pF

	Class of Product			
	A	**B**	**C_1**	**C_2**
Up to V_1: No error or malfunction at all	NA	4kV	3kV	5kV
Up to V_2: Only recoverable errors permitted	10kV	15kV	10kV	22kV
V_3: Hard errors, but no solid damage nor unsafe conditions created (for example, unexpected re-start of a machine, an elevator, etc.)	20kV	25kV	25kV	30kV

*Note: The categories in this table are the same as those in Table 4.2. The same voltage values as for the "personnel-only" discharge of Table 4.2 have been kept. This seems like overkill since the IEC discharge network representing only 250 ohms (with short ground strap) the current will be several times higher than for personnel. However, the IEC-type rise time is also much slower, which compensates for this, while the breakdown properties of a high voltage are kept. On the other hand, although the oscillatory waveform of real furniture ESD is not simulated, the large current and longer time tends to produce a somewhat similar effect. *In any case, there is no way a single discharge network can perfectly replicate both personnel and furniture ESD.*

4.5 The Error per Discharge Concept

Not exclusive of, but rather complementary to the severity levels, another concept has been used recently. A good rationale of this is explained in Ref. 24. The concept is based on the fact that the error sensitivity of a machine to ESD is almost never a step function. Fig. 4.15 shows conceptually an "ideal" behavior; i.e., a machine experiencing no error at all (regardless of how many pulses are applied) below a given ESD level, then experiencing one error per pulse or (P) error per discharge=100% above that level.

By comparison, the behavior of an actual machine is plotted as the number of errors per ESD event, which is less than or equal to unity, since in fact:

$$\text{No. of errors/ESD pulse} = \frac{1}{\text{No. of ESD pulses to cause an error}}$$

This number is recorded versus the ESD voltage of the simulator.

Due to the random occurrence of the ESD event versus the operations of the EUT, the error/pulse concept allows the replacement of the GO/NO-GO concept with a better approach. The assumption is that the product designer can establish the maximum number of errors per day (or per shift) that his product—and its user—can tolerate.

Considering that, in field conditions:

$$\text{No. of errors/Shift} = [P\ (\text{Error, V}) \times N\ (\text{Event} \geqslant V)\ \text{per shift}]$$

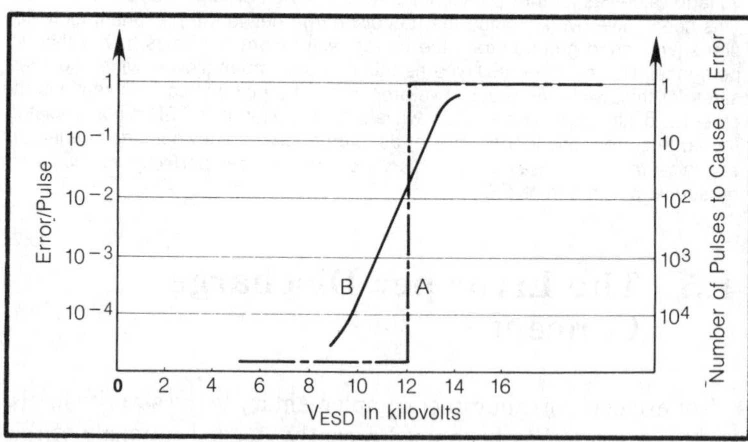

Figure 4.15—The concept of error per pulse
Problem: The behavior of a machine during ESD test is not a step function as in (A) but a steep slope like (B) due to the random occurrence of ESD pulses vs. logic operations.

where,

P (Error, V) = Probability that the machine will make an error, given a discharge voltage V

N (Event≥V) = Number of ESD events which will equal or exceed V

One can derive that:

$$\text{Error rate criteria at a given test level } V_{ESD} = \frac{\text{Tolerable No. of Errors/Shift}}{\text{No. of Events/Shift exceeding V}}$$

However, this equation does not consider the slope of the error curve. The probability of ESD events *decreases* continuously with increasing voltages. Inversely, the probability of making one error per pulse *increases* with increasing voltages. An example will show how the error rate can be handled: Fig. 4.16 shows a histogram of personnel ESD for the worst-case months (RH=15-20%) compiled from Simonic's data (see Sec. 1.4.1). Although the ESD events were given in amps, they have been translated into an equivalent body voltage assuming a uniform discharge impedance of 1000Ω. This will allow obtaining the results directly in testing voltage.

Assume that the objective is to have less than 1 error/shift (5 errors/week) during the worst-case months and for the worst-case installations (synthetic carpet). The EUT error profile, in its initial stage, has been plotted on curve A. For 7 kV, with a simulator set for personnel (about 1000 Ω of discharge resistance), there is one error per shot. For 4 kV it takes, on an average, 30 pulses to cause one error.

For each 2 kV interval of the histogram, the corresponding error/pulse figure is multiplied by the number of events per shift to come up with the number of errors per shift. The total amounts to 2 errors/shift, which is excessive.

In general, the slope of the error curve has nothing to do with the fixes (shields, decouplings, etc.) existing in the machine, but rather with the nature of the operations and messages performed by the EUT. Therefore, hardware improvements will not change this slope, but merely shift the curve. For instance, curve B shows the same unit after certain ESD fixes. The total is now 0.65 error/shift, which is below the one error objective.

An ultimate refinement, described in Ref. 24, is to assign a weight (0 to 100%) to each side or zone of the machine to represent the percentage of ESD events which will occur at that position. The sum of all the weights over all test points is 100%. Therefore, parts of the machine which are less exposed than others will be given a lesser penalty.

The merit of the referenced study lies in the fact that it replaces the traditional single GO/NO-GO voltage* with a finer characterization of the machine response. However, the search for machine error rate and position weighting can be delicate and is not the kind of iteration which can be made during manufacturing QC testing. Therefore, the following is recommended:

a) At the end of the design stage, when the prototype is sufficiently representative of the final product, especially concern-

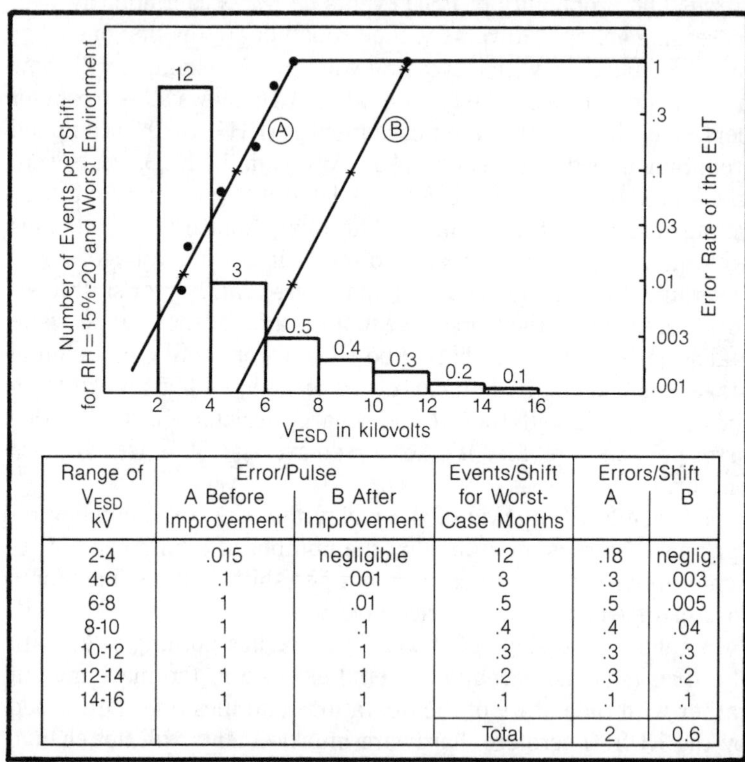

Range of V_{ESD} kV	Error/Pulse A Before Improvement	Error/Pulse B After Improvement	Events/Shift for Worst-Case Months	Errors/Shift A	Errors/Shift B
2-4	.015	negligible	12	.18	neglig.
4-6	.1	.001	3	.3	.003
6-8	1	.01	.5	.5	.005
8-10	1	.1	.4	.4	.04
10-12	1	1	.3	.3	.3
12-14	1	1	.2	.3	.2
14-16	1	1	.1	.1	.1
			Total	2	0.6

Figure 4.16—Example of application of the error/pulse concept

*The reader sees that the number of discharges given in existing specifications are extremely arbitrary. As long as the error curve of the machine is not known, imposing a number of discharges based on a general standard is about as sound as picking that number by throwing dice.

ing the logic speeds, clock rates, architecture and sequence of the logic functions, type of microprocessor, type of interface protocols, etc., a characterization of the machine ESD error curve should be made comprising:

- Machine error rate over the complete voltage range, by 1 or 2 kV increments.
- Position weighting
- ESD environmental weighting of (controlled or uncontrolled) RH, etc.

b) Knowing the above, engineering tests will be pursued to see if the machine meets the maximum error/shift or error/month objective, and improvements will be made if necessary.

c) From then on, the routine QC test will not need to repeat the above, but will merely check the average error rate at one voltage only. For instance, in our example, if the simulator is set to 10 kV, the QC will ensure that it takes, on the average, a minimum of 4 pulses to cause one error. As an added security, a sample check could be done to verify that, at 7 kV, the machine makes less than 1 error per 100 pulses (this, according to the IEC criteria, would go undetected). This will allow sample comparison and detect possible degradation after hardware changes. The slope of the error curve will not change, but its position will shift.

4.6 ESD Test During Design and Development

Given that ESD testing is a very powerful, versatile, and relatively easy test to conduct, it is recommended that it be applied to a machine as soon as an early prototype exists and, furthermore, *as soon as functional sub-assemblies* exist. For instance, an early ESD test is easy to perform on a breadboard prototype using an indirect discharge test set-up. One of the most rewarding approaches is to start an ESD evaluation at the PCB level. When functional cards have been designed and the first run of prototype cards have been assembled, the following method is recommended:

- Identify the principal cards in the machine; i.e., those which perform essential functions and constitute block-diagrams in the machine architecture.

- Prepare each of those cards so that they can be tested as stand-alone items. For instance, identify a few lines to be considered "witnesses" of the card's good condition (for example, reset lines in a microprocessor) and equip them with an LED soldered directly onto the card, so that when everything is normal, the LED is either ON or OFF. Also provide a dc voltage without the need of an external bench supply. The easiest way to do this is to merely attach an ordinary battery pack to the card with adhesive tape and solder the connections to the supply bus via ultra short, twisted wires.
- Perform an indirect ESD test of the card as shown in Fig 4.17. The discharge is made on a metal plane, the card being placed at a distance which is representative of the actual card-to-housing distance of the future machine. The test voltage depends on the criteria for the final product. If the product

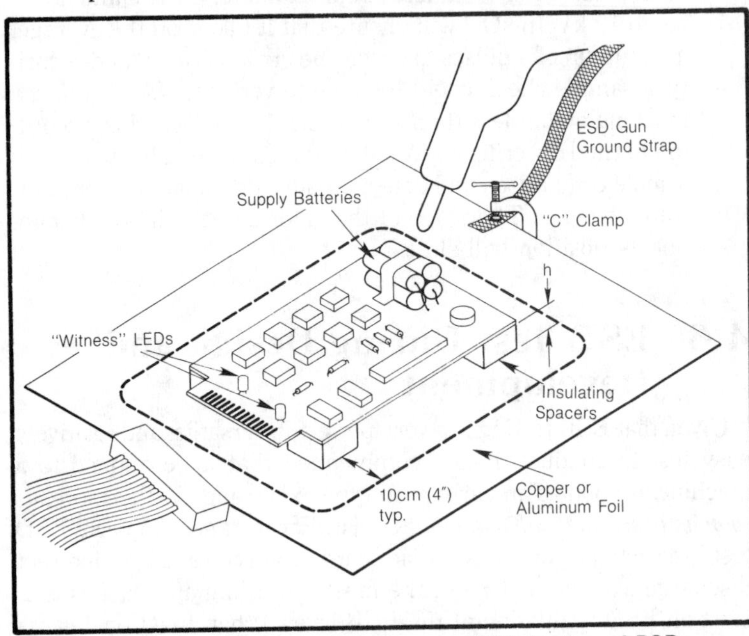

Figure 4.17—Workbench mounting for early ESD testing of PCBs. The height "h" depends on the final configuration of the machine; i.e., the average distance to card will be from the bottom plate or the closest conductive wall of the housing. By default, 5 cm (2") can be used. The cable will be used later to address the susceptibility of the interface. The discharges are made following a route parallel to the card perimeter. One point of discharge at each corner, plus one midway, is usually sufficient.

is planned with a plastic, non-conductive cabinet, the ESD voltage should be set as for the final product. If the machine will have a metal or metallized cabinet, a certain derating should be taken. This derating could be found by testing a similar machine (identical size, identical cards location) with and without covers; the difference in kilovolts between the two conditions is the derating. By default, a good rule of thumb is to test each stand-alone card for 6-8 kV. A machine equipped with cards which sustain that level will not be too difficult to harden up to 15 kV or more.

This single card test probably represents some of the best invested time in the entire ESD strategy (perhaps even in the entire EMC strategy). It reveals all the PCB layout weaknesses (see Chapter 5) at a time when they are relatively easy to correct. By using the set-up of Fig. 4.17, it will be quite simple to identify the weak spots and find the peculiarity of the layout in that area which causes that weakness.

- When the card has been hardened to the desired level, one thing remains to be done: hardening the input-output lines. A well-hardened card can still make errors if ESD-induced glitches arrive by the connector pins (remember, the test done so far involved only the direct radiation pick-up by the card). To this end, a typical length of flat cable, multipair cable, or anything which replicates reality should be connected to the card connector(s). Due to the localized nature of ESD, it is generally not necessary to put more than one or two meters of cable (at one meter, the cable is already beyond $\lambda/2$ of the typical ESD bandwidth). The cable, on its other end, should be terminated with passive resistances simulating the actual impedance seen at this end. The test could be refined by connecting this cable to an exerciser which will actually reproduce the normal I/O transactions with the card. However, great caution must be taken in that case to be sure that the part of the gear which is stressed is actually the card, and not the exerciser.

4.7 ESD for Field Diagnostics and Forced Crash

Until the early 1980's, it seemed like electromagnetic interference and electrostatic discharge were two different worlds. The two worlds paralleled, and in many companies and organizations, the same people were often wearing two hats . . . but these were still *two hats;* the issues were different. EMI susceptibility studies and tests were aimed at the behavior of equipment exposed to stationary electromagnetic ambient fields and to power line disturbances, while ESD tests attempted to reproduce human body or furniture discharges and the effects of such discharges on fragile integrated circuits.

It is known from transfer function theory that any system can be viewed as a black box with input/output ports. To the degree a system can be approximated as a two-port network, its transfer function can be determined very conveniently by measuring its response to an input impulse. In some five years of extensive experience solving ESD problems, there were many times when the author had to use a small coin as a screwdriver because no screwdriver was available; likewise, the convenient, easily carried ESD simulator can be used as a first, rough check of susceptibility to almost *any kind of EMI.*

It is clear that a properly simulated ESD event can excite and reveal much more of the weaknesses of a product than its mere vulnerability to static discharges.

Testing radiated EMI susceptibility by the traditional method is a long and complex process requiring a costly set of amplifiers, antennas and anechoic rooms. Although the ESD test cannot pretend to replace the traditional method, the wideband field which flashes the product all at once reveals many of the same weak spots that classical EMI tests explore. Even more, an ESD test is usually very simple to run and is often ideally suited to field testing. It will help fix field problems by using the *forced crash tactic,* giving the sick machine an ESD "vaccination."

Many computers and microprocessor-based equipment experience random malfunctions once they are installed, even though they were declared *good for shipment* by the manufacturer's quality control people. These environmental problems quickly become exasperating because of their typically intermittent and inconsistent nature. Usually after the first customer call, a field technician checks out the machine, trims some settings, replaces a couple of circuit cards and leaves. (Needless to say, no malfunction occurs while the field technician is there to see it!) Back at the office, the technician, in good faith, will probably report a "NTF" (no trouble found). Of course, the problem recurs at the same unpredictable rate, correlated to nothing or to so many things that a sound explanation seems impossible. The customer asks for a higher level of assistance and the district field engineer comes to the rescue, and so on, until someone eventually calls an EMC specialist.

Is the problem ESD? Well, maybe or maybe not. In any case, an ESD test applied on site will be merciless in pinpointing:

- improper high frequency grounding of computer system
- deficient subsystems attached to an otherwise healthy system
- poorly shielded I/O cables, or original manufacturer's cables and connectors replaced by *look-alike* (but *not* perform-alike) items.
- missing covers, gaskets or fingerstocks
- covers improperly bonded to mainframe, etc.

The recommended ESD test has two variations. The simplest variation is as follows:

1. Using a piece of chalk, etc., make a *zoning* and coding of the EUT housing (such as that decribed in Sec. 4.3).
2. Establish a clear, indisputable "fail" status of the EUT.
3. Set an ESD level equal to (or correlatable to) the one used by the manufacturer's Quality Assurance, if one is specified. Otherwise, by default, select 10 kV for a simulator like the Schaffner or KeyTek which normally has a 150 ohm/150 pF discharge head. *To ensure good repeatability, stabilize the ESD return path by installing a temporary aluminum foil over the floor underneath the machine and grounding the ESD gun to it.*
4. Zap every coded area, including every switch, key, protruding screw, etc., with at least 50 pulses.
5. If the product is mainly plastic, simulate an indirect discharge by zapping a metal plate about 4″ (10 cm) from each side.
6. Record and map the *pass* and *fail* areas.
7. For each failed area, lower the test voltage until reaching the GO/NO-GO level, and document the results as suggested in Fig. 4.13.

Note that the above test requires no equipment other than the ESD simulator.

For the test's second variation, the use of an oscilloscope or spectrum analyzer is required, and the ESD simulator is used in the *continuous* mode. This allows a user's selected rate of ESD pulses to which the measurement equipment can be synchronized. This second variation should be used when a failure analysis indicates that the system under test has the ability to self-correct any ESD induced malfunction in between ESD pulses in such a way as to deny the tester a convenient failure indication. It is vital when using any test gear to be sure it does not corrupt the test. Opening up the equipment under test to connect oscilloscope leads is a NO-NO! Placing a current probe on a line in a separate interconnecting system is a likely safe measurement point.

Once repeatable pass/fail measurement criteria have been confidently determined, what remains is to inspect and critique the hardware immediately behind each failing zone, and especially nearby cables. Be on the lookout for I/O cables or power cords. They are privileged points of ESD entry because they act as efficient pick-up antennas for the ESD field existing outside. Then the induced current is carried inside by shield pigtails, unprotected wires, etc. For I/O cables, a radical improvement is obtained when ordinary or poorly shielded cables are replaced by homogeneous shields plus metallic connector shells bonded to the main frame. If an ESD weak spot exists near the power cords, a close look will probably show one of these typical mistakes, all detrimental to machine immunity:

- poor filter, or a filter improperly bonded to the chassis
- filter located too far inside
- input and output wires of the filter tangled
- ground wire extending too far inside the machine.

Once changes have been applied to the failing areas, re-run a test to see if they now meet or exceed the pass/fail level. Be sure to re-run the test on all zones, even those which were previously okay: ESD and Murphy are old mates, and a local *improvement* may have caused a degradation in another area (especially any improvements requiring rerouting of wires)!

The philosophy behind all this is that *if a unit, at its site, with all external cables and peripherals attached, is fixed to 10 kVolts or more, it will be very hardened to any type of EMI, whether it is ESD or not.*

There are, of course, exceptions to our recipe. First, if the trouble was a casual power line undervoltage or overvoltage during a half cycle or so, an ESD check will not detect it. But these troubles are usually easy to trace by putting a *spy* power line monitor on the suspected outlets.

Remember, ESD is a single isolated event. More and more systems are fault tolerant. The ESD-induced error can be self-corrected by parity or cyclic redundancy checks, etc., and a software recovery can make the error *transparent* to the user. If this is suspected, the ESD simulator should be used *continuously* with a nonperturbing measurement system synchronized to the ESD pulse repetition frequency. For typical civilian environments and many military environments, these ESD simulator EMI tests in the field should reveal most system EMI deficiencies. In those cases where it won't, the problem is probably due to a *continuous* ambient field like a nearby radio or radar transmitter. However, (except for walkie-talkies) such a situation is generally not *random* and is usually detected by a quick ambient field survey.

Chapter 5

Design for ESD Immunity

ESD should not be fought with twelfth-hour fixes and costly retrofits. Rather, ESD should be treated as *any* potential environmental condition; that is, it should be dealt with during the *design* of an equipment. There are many cases where ESD immunity has been built in at the *PCB* and *INTERFACE* levels so that even without additional shielding of the cabinet, the equipment can withstand ESD levels above 10 kV.

ESD immunity can be considered at the following stages:

- at component level
- at circuit board level
- by software and noise inhibition features
- at internal packaging and wiring level
- at housing/cabinet level
- at external cabling level
- at installation and environment level

Full ESD protection can be implemented at one of these levels only. Costwise, however, it is generally more efficient for ESD control to be shared among several levels.

5.1 ESD Protection at Component Level

A first degree of ESD immunity can be achieved by selecting components (logic ICs, analog ICs, operational amplifiers, resistor networks, etc.) that already have built-in ESD protection. Appendix A gives some information on the types of static damage protection features incorporated by some vendors.

It is cost-effective to purchase ESD immune components (for instance, components with a 2 kV to 4 kV ESD grade) rather than adding transient protection devices outside each component. However, these integrated Zeners and guard rings are efficient against ESD damages, but they cannot prevent errors if a transient of a few volts is induced by ESD. Also, even a 4 kV hardened chip is still vulnerable if it is directly wired to a switch, keyboard, etc., that can receive a 15 or 20 kV direct ESD.

If no technology or vendor can be found with built-in ESD protection that matches the objective, external protection can be added close to the most critical pins by using:

A) Discrete or integrated transient protectors like the ones shown in Fig. 5.1. These devices have been designed to minimize parasitic inductance. The dual-in-line package of Fig 5.1a uses a "forked" leads approach to prevent lead inductance from appearing across the line to be protected. The device of Fig 5.1b uses the new surface mounting method.

B) Series resistances on the sensitive, high impedance inputs to limit the current below damage level.

C) Capacitive decoupling near the critical pins. For instance, on the power input pin, a ceramic cap of a few nanofarads can be added even though it was not considered necessary during the initial circuit design.

An interesting alternative is shown in Fig. 5.1c where the decoupling capacitor is integrated in the substrate of the IC to act right at the supply pins level.

On signal inputs, 100 pF or more can be added, depending on what this line can tolerate as far as capacitive loading without affecting the performance.

While the transient protectors will not eliminate low level glitches (those above the detection threshold but not high enough to cause damage), decoupling capacitors will. However, alternatives A), B) and C) have their limitations:

- they rapidly add to the parts count and to the cost, a disadvantage if these aspects are critical.
- they work against the integrated circuit speed and performances. Varistors and Zeners, especially, have parasitic capacitances exceeding 100 pF.

To protect extremely low level inputs, such as a few hundred millivolts or less, another solution is to use a low capacitance signal

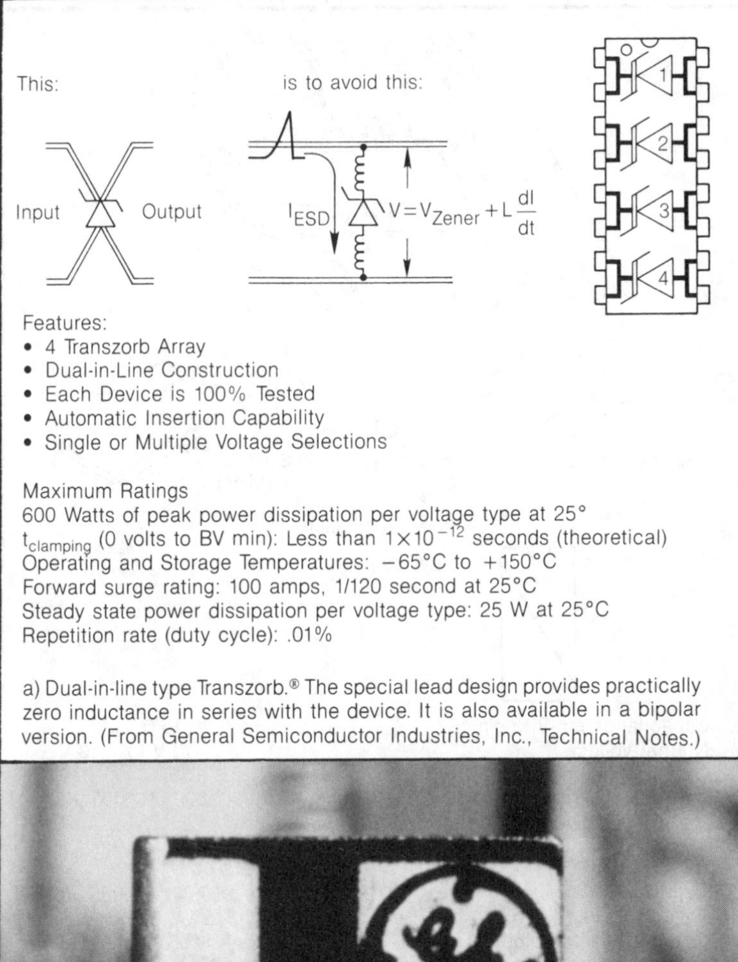

a) Dual-in-line type Transzorb.® The special lead design provides practically zero inductance in series with the device. It is also available in a bipolar version. (From General Semiconductor Industries, Inc., Technical Notes.)

Courtesy of General Electric

b) Surface-mount metal oxide varistor. The ... ar protection.

Figure 5.1a—Integrated transient protectors

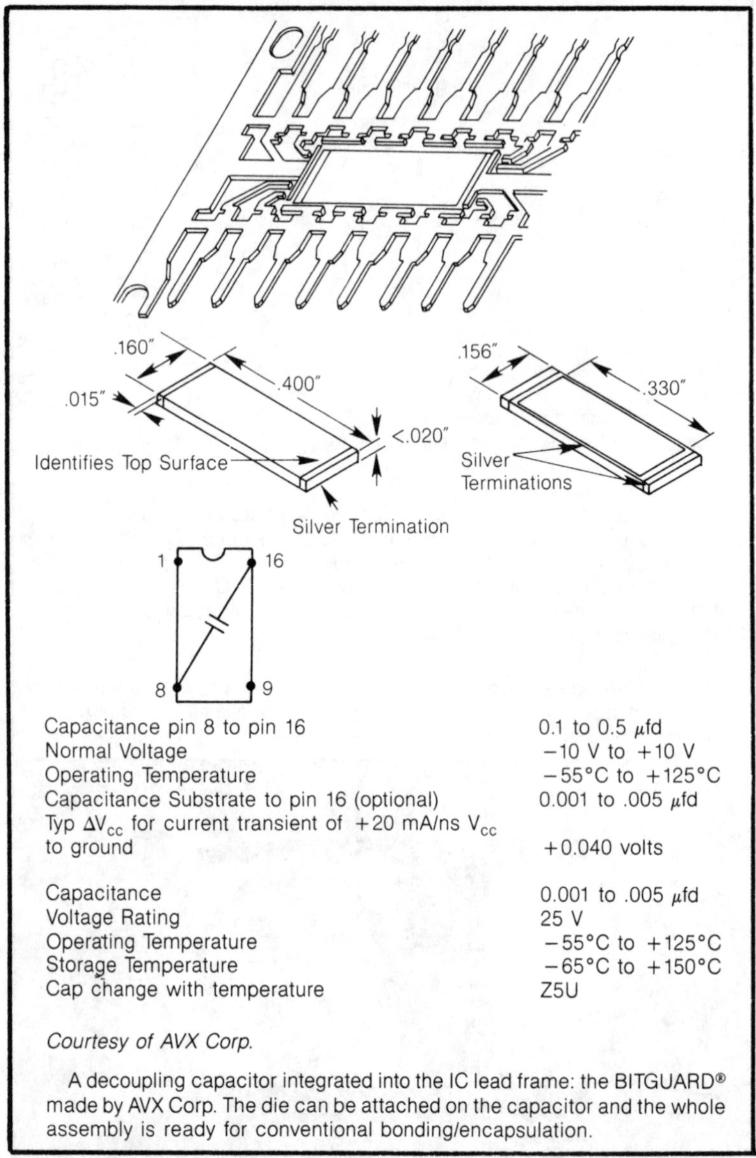

Capacitance pin 8 to pin 16	0.1 to 0.5 μfd
Normal Voltage	−10 V to +10 V
Operating Temperature	−55°C to +125°C
Capacitance Substrate to pin 16 (optional)	0.001 to .005 μfd
Typ ΔV_{cc} for current transient of +20 mA/ns V_{cc} to ground	+0.040 volts
Capacitance	0.001 to .005 μfd
Voltage Rating	25 V
Operating Temperature	−55°C to +125°C
Storage Temperature	−65°C to +150°C
Cap change with temperature	Z5U

Courtesy of AVX Corp.

A decoupling capacitor integrated into the IC lead frame: the BITGUARD® made by AVX Corp. The die can be attached on the capacitor and the whole assembly is ready for conventional bonding/encapsulation.

Figure 5.1b—Integrated transient protectors

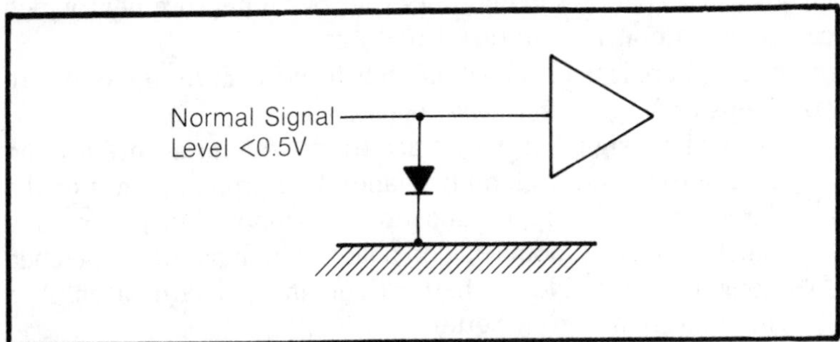

Figure 5.2—Protection scheme for very low level inputs. If one wants to also protect from negative transients (i.e., coming from the ground), another diode can be added.

diode directly polarized as in Fig. 5.2. The diode must be rated for the peak forward current corresponding to a worst-case ESD condition; for example, 20 Amperes during 100 nanoseconds. Since a silicon PN junction has a forward barrier voltage of about 0.6 volts, the diode will show up as an infinite resistance for the normal signal, but will be conductive for voltages above 0.6 volts.

A third approach, too often overlooked, is *to be sure not to use a too fast technology unless it is needed for a specific function.* Many times, logic families with speeds of 3 or 5 nanoseconds are used all over the board by a sort of intellectual license assuming that "the faster, the better." Chapter 2 showed that one can achieve good immunity, at no additional cost, by using devices with the lowest bandwidth compatible with the function. Nevertheless, the designer must watch for the input impedance of the device. A device ten times slower will not show much improvement over a TTL or STTL if it has a quasi-infinite input impedance. Here, pull-up/pull-down resistors can help to decrease the vulnerability of high impedance inputs.

5.2 ESD Protection at the PCB Level

The printed circuit board is probably the area where the improvement/cost ratio is the largest. The effort invested in sound PCB layout will be paid for many times over by an improvement which can be drastic at no additional parts cost.

The rule here is to check all the runs to avoid large signal-to-zero volt loops or V_{supply}-to-zero-volt loops:

- Never let a signal or a V_{cc} trace travel without a close ground return (trace or, preferably, plane). Use ground planes or the largest possible copper lands to act as a noiseless ground and shield. Similarly, double-sided boards with maximum unetched copper on one side are better than single-sided boards. And multilayers are even better.

  Fig 5.3 shows several "no-nos" in PCB design. Once *all* these loops have been purged from the layout, the ESD immunity

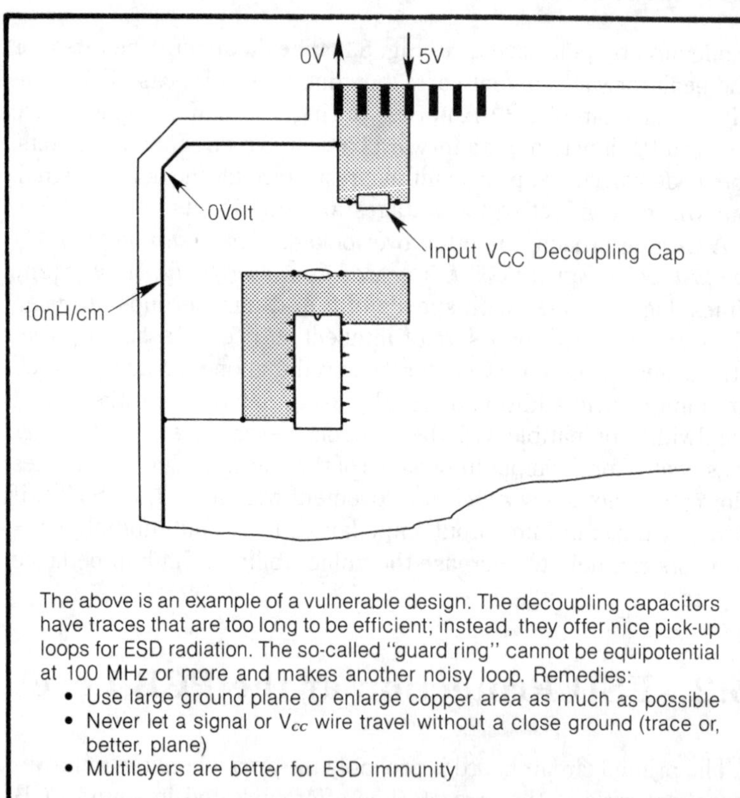

The above is an example of a vulnerable design. The decoupling capacitors have traces that are too long to be efficient; instead, they offer nice pick-up loops for ESD radiation. The so-called "guard ring" cannot be equipotential at 100 MHz or more and makes another noisy loop. Remedies:

- Use large ground plane or enlarge copper area as much as possible
- Never let a signal or V_{cc} wire travel without a close ground (trace or, better, plane)
- Multilayers are better for ESD immunity

Figure 5.3—ESD protection "by design" starts at the PCB level

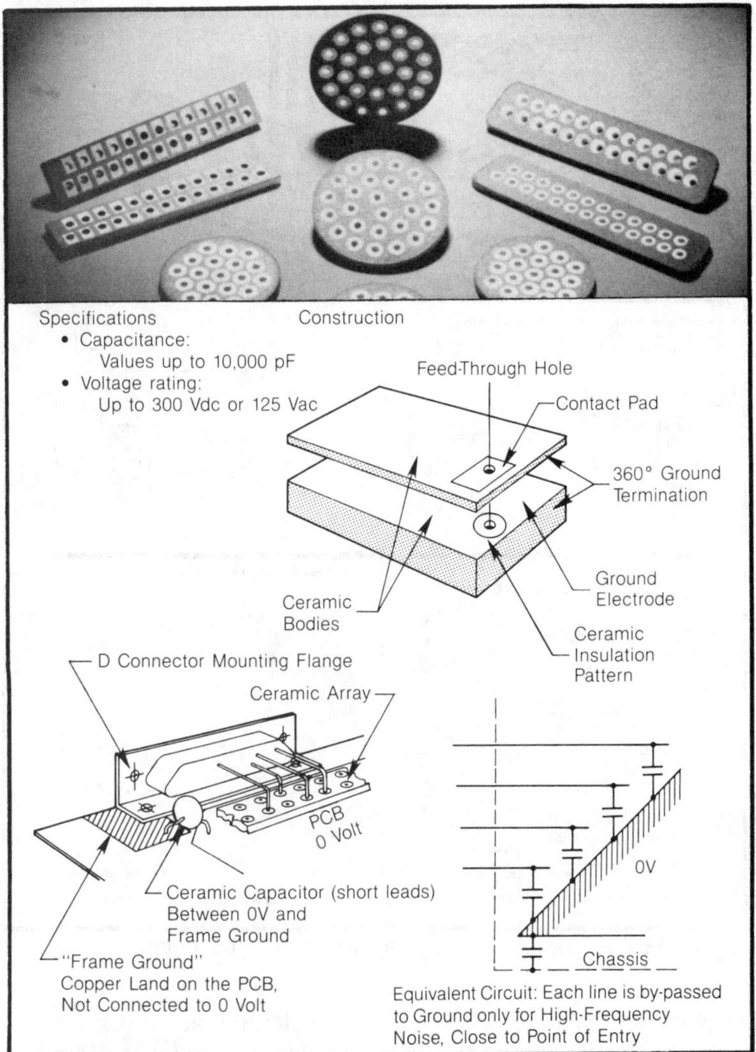

Figure 5.4—Planar capacitor arrays and their mounting. The discrete capacitor between the 0V and chassis is to avoid low-frequency ground loops while still providing a direct sink for fast ESD transients. Another approach is to provide a copper land around the capacitor array, and to bond this "guard ring" directly to chassis.

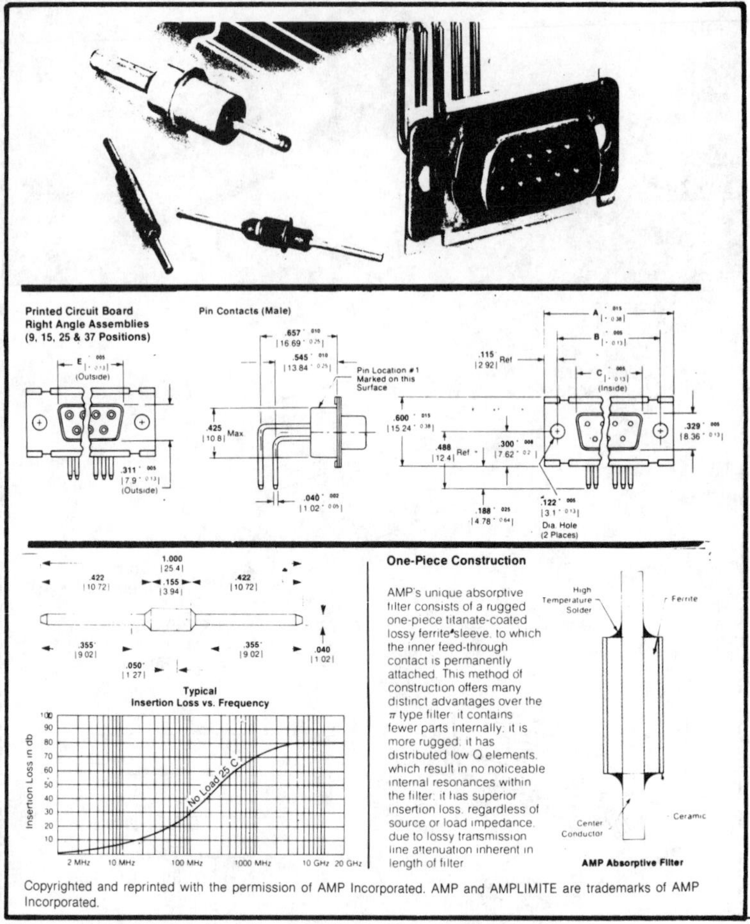

Figure 5.5—'20' series filtered AMP "Amplimite" connectors

of the machine will improve considerably at no cost.

- Eventually, use a compartment shield to protect critical areas where a ground plane cannot be implemented or is insufficient. This shield should be most directly grounded to chassis to prevent the shield from re-radiating on the PCB or contaminating the 0volt with shield currents.

Special attention must also be given to connector areas, especially those connectors receiving I/O cables directly from the outside. Nor-

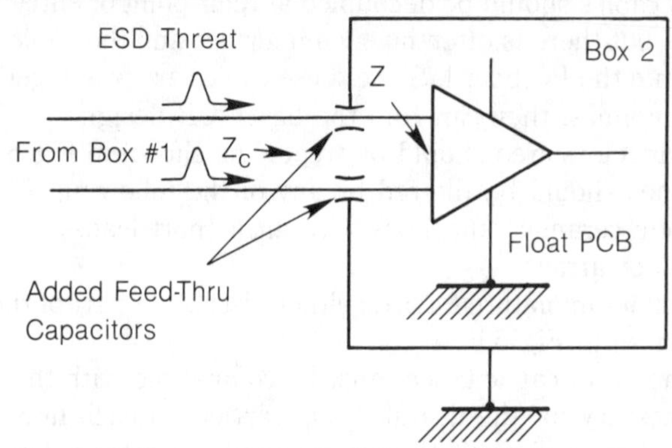

To avoid compromising wanted signal by feed-thru capacitors:

$$\frac{1}{2\pi f_2 \frac{C}{2}} \geqslant 3Z \rightarrow \text{good pulse integrity}$$

$$\frac{1}{2\pi f_2 \frac{C}{2}} \geqslant Z \rightarrow \text{marginal pulse integrity}$$

f_2 = 2nd corner frequency of wanted spectrum

$$= \frac{1}{\pi \tau_r}$$

Z = Differential impedance (source and load in //)

	Slow Speed Interface (typ.)	CMOS ckt	TTL ckt
Signal Rise Time, τ_r	1μs	100ns	10ns
Bandwidth	300kHz	3MHz	30MHz
Z	100Ω	500Ω	100Ω
C_{Max} for good pulse integrity	3000pF	60pF	30pF
C_{Max} for marginal pulse integrity	10,000pF	200pF	100pF

Figure 5.6—Small compromise to floating scheme to decouple ESD and ultra-short CM transients

mally, I/O cables should be decoupled at their point of entry in the enclosure, but there is often no such interface and the cables plug directly onto the PC board. Since these cables are privileged ESD pick-up antennas, they can ruin the best PCB design.

The connectors area should be treated as shown in Fig 5.4. Incoming lines should be filtered by any of the following:

- discrete ceramic capacitors with ultra short leads
- capacitor arrays
- capacitor+miniature ferrites, either discrete or part of the connector as in Fig 5.5.

The amount of capacitance must be compatible with the bandwidth necessary for the useful signal carried on each line. Since these capacitors will preferably be mounted in a common-mode arrangement, if a capacitor C is put on each line to chassis, the line will see a capacitance C/2 across it.

To get the most benefit from these decoupling devices, follow these simple rules:

- Compute the maximum capacitance tolerable without signal distortion (see Fig 5.6).
- If ferrites are also used, have the ferrite looking towards the signal source (generally low impedance) and the capacitor looking towards the load (generally high impedance).
- Connect all the ground terminals of these capacitors (or arrays) to a copper land surrounding the connector area. This land should preferably be distinct from the 0volt plane and connected with a short strap to the nearest chassis point. Thus, the ESD currents diverted by this filtering will go to the frame ground with the least possible disturbance on the card.

5.3 ESD Immunity by Software and Noise Inhibition Features

Too often neglected, "intelligent filtering" by signal recognition and software can be used efficiently by:

- AND gating of critical signals with an *undisturbed* clock line
- Use of parity check, CRC checks, auto-correlation, etc., techniques
- Use of "graceful recovery" schemes.

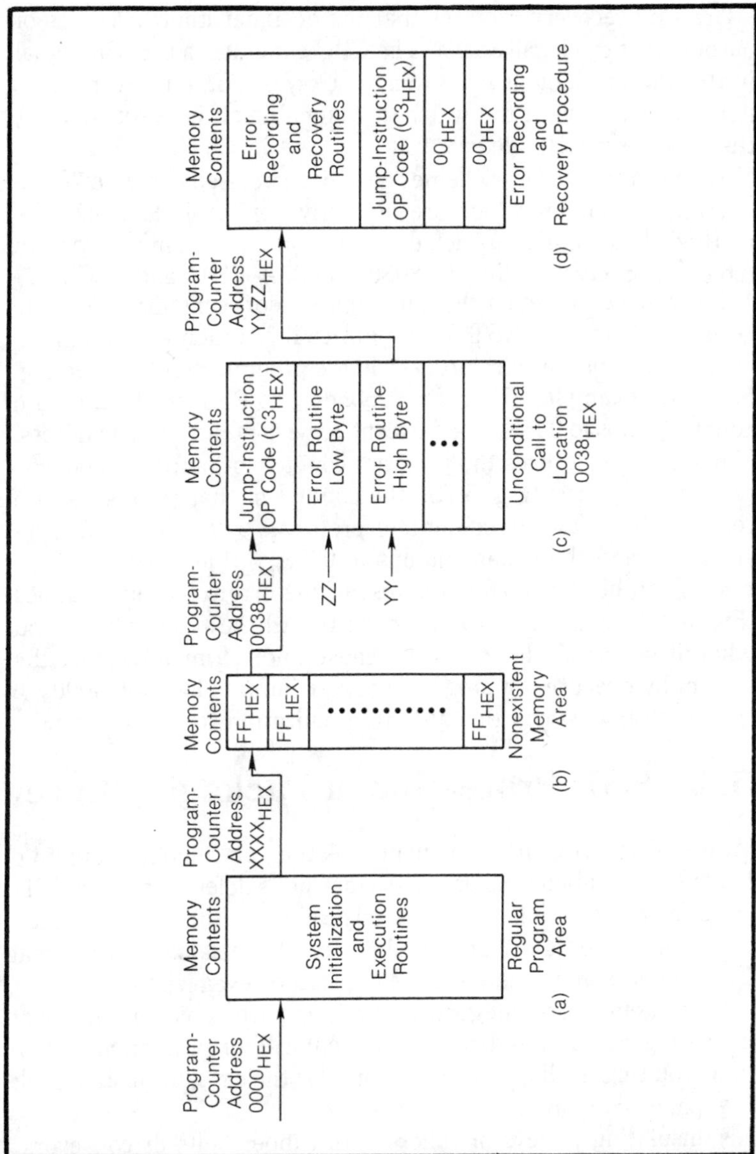

Figure 5.7—Program-counter sequence shows how software routines can be used to allow a microprocessor to regain control after program-counter jumps to nonexistent memory areas occur

Graceful recovery requires that the designer know all kinds of outputs that can legally exist. The CPU generates a periodic signal (watch-dog) as long as everything is okay. If not, a recovery is initiated. Of course, if the watch-dog signal itself is upset by ESD, the scheme may be defeated.

Many error-recovery schemes have been devised to create "fault-tolerant" machines which are perfectly applicable to ESD. The method shown in Fig 5.7 achieves automatic program recovery on 8-bit microprocessors like the 8080A, 8085A, Z80A and NSC800.*

The instruction set for these microprocessors contains single-byte restart calls labeled RST 0 through RST 7. When a program encounters one of these restart calls in the program memory (see Fig. 5.7a)—for example, FF_{HEX}, corresponding to RST 7—the program counter jumps to 0038_{HEX} and executes the contents of that address. When an error causes the program counter to jump to a nonexistent memory area (Fig. 5.7b), the content of that address is also read as FF_{HEX}. The processor interprets this byte as the RST 7 instruction, and the counter jumps to 0038_{HEX} (Fig. 5.7c).

This can be the starting address of the error-recovery routines (Fig. 5.7d); then, those routines can record the type of error and when it occurred, determine its cause and, ultimately, reset the system by executing a program-counter jump to location 0000_{HEX}, the location at which the system initialization routines begin.

5.4 ESD Protection at Packaging Level

Several practices in component selection, placement, mounting, bonding, and cabling can upgrade a machine's defenses against ESD.

As far as selection, there are basically two things to consider. Some components or subassemblies can be *generators* of internal ESD, while some others can be carriers or even victims.

In the generators category, we find the parts which are made of insulating material and can be momentarily or permanently rubbing, rotating, sliding, etc. This would be the case, for instance, of:

- paper containers
- insulating pulleys or rollers with rubber belts or conveyors
- cooling fans with polycarbonate blades

When these components are selected, preference should be given to those which the manufacturer has designed as anti-static or static-free through material selection, conductive additives, or shielding.

*This automatic program recovery method is reprinted from EDN, December 17, 1984. ©1984 Cahners Publishing Company.

All metal parts around these products should be grounded to the metallic mainframe of the machine. For sliding or rotating parts, wiping contacts or conductive grease can be used. However, these solutions may create graphite dust or maintenance problems (Fig. 5.8).

If regular maintenance cannot be counted on, smooth wiping with soft, conductive, grounded brushes can be used (Fig. 5.9). As a varia-

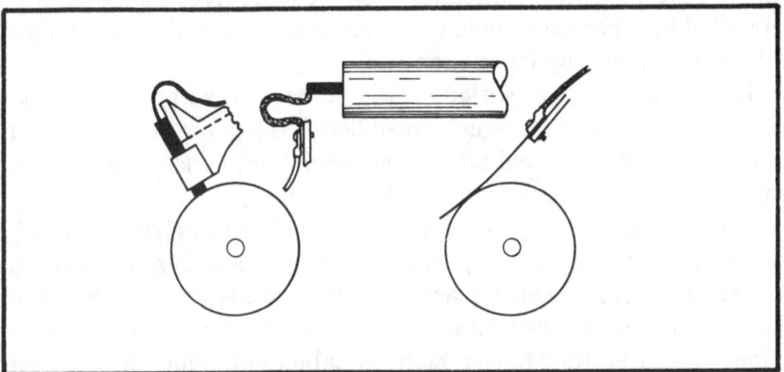

Figure 5.8—Grounding of shafts and rotating parts: Carbon brush or bronze spring

Courtesy of Chapman Corp.

Conductive tinsel is a low-cost temporary solution. It is installed in contact with sheet surfaces and conducts the static charge to ground. When it loses its shape, it must be replaced. A static brush is used for a longer lasting solution.

Spring tinsel devices can be designed for specific applications. Easy to use and install, they are a low-cost solution for minor ESD problems. The device on the left is composed of two stainless steel springs and two lugs. The device on the right is a solid brass spring with copper tinsel drawn through it.

Figure 5.9—Tinsel and spring tinsel devices

tion of this, metallic tinsel devices arranged as antistatic combs or neutralizers have been developed. In this case, the multitude of grounded needles act as a static "collector" which bleeds the charges to ground even without physically touching the electrostatically charged object.

High voltage dc supplies, due to the non-null value of their static field, can charge all insulated parts in the path of the electric field lines. To reduce this charge, high-voltage dc sources should be surrounded by a Faraday shield to minimize their parasitic capacitance to the other nearby components.

In extreme cases of serious static generation within a machine, active static eliminators can be installed. They generate a local high voltage alternating field which ionizes the air, hence allowing the opposite charges to recombine (Fig. 5.10).

The second category of components, the "carriers" or "victims," is by far the largest. A special problem is posed by keyboards, especially membrane keyboards. These devices cause a large opening in the covers, creating a privileged entry port for ESD radiation. Moreover, they have a high probability of being hit by direct discharge, with the discharge arc directly reaching the sensor circuits under the domes or capacitive arrays (Fig. 5.11a). An example of an ESD shielded membrane keyboard is shown in Fig. 5.11b. Note that an interlayer of thin steel has been provided and is multipoint grounded to the chassis so that the discharge current does not penetrate the inner part of the keyboard. Cheaper versions exist where, to save one layer, the flexible contact layer simply includes a grid of conductive ink claimed to be a "shield." Although this "shield" reduces radiation, it is not efficient enough to block direct ESD.

Finally, ESD has to be considered when it comes to the location of sensitive items like PC boards, magnetic disks and their logic, etc. Not only should the proximity of the ESD sources described above be avoided, but also the proximity of possible ESD routes such as cooling or display holes, seams, openings, etc. Near these areas, the shielding effectiveness of the housing will be null or minimal, and ESD currents will re-radiate inside. If covers are bonded by straps, be aware that, during a discharge, these straps will generate a strong magnetic field. Therefore, no sensitive components or their cabling should be placed close to them.

Cables deserve special mention. *Internal wiring,* being exposed to ESD re-radiation inside, will carry some induced ESD noise. Besides the precautions at their termination into the PCB connectors, the following guidelines will prevent excessive pick-up:

- avoid long runs of cables along cover seams, hinges, and bonding wires

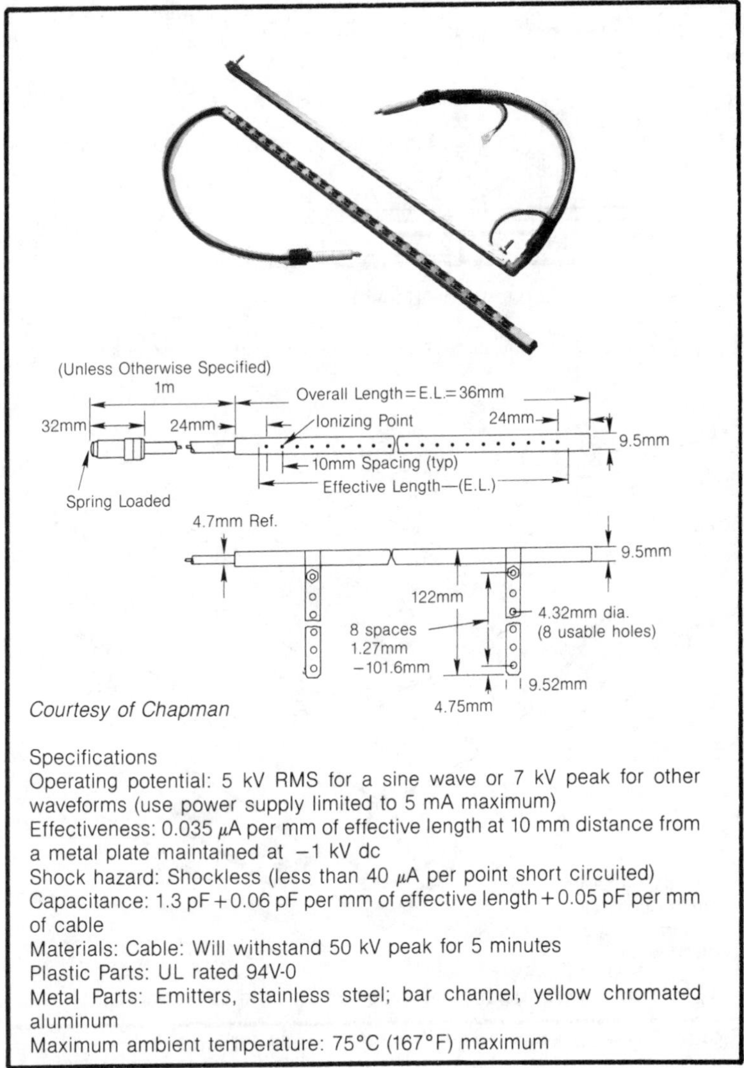
(Unless Otherwise Specified)
1m
Overall Length=E.L.=36mm
32mm
24mm
Ionizing Point
24mm
9.5mm
10mm Spacing (typ)
Effective Length—(E.L.)
Spring Loaded
4.7mm Ref.
9.5mm
122mm
4.32mm dia.
(8 usable holes)
8 spaces
1.27mm
−101.6mm
9.52mm
4.75mm
Courtesy of Chapman
Specifications
Operating potential: 5 kV RMS for a sine wave or 7 kV peak for other waveforms (use power supply limited to 5 mA maximum)
Effectiveness: 0.035 μA per mm of effective length at 10 mm distance from a metal plate maintained at −1 kV dc
Shock hazard: Shockless (less than 40 μA per point short circuited)
Capacitance: 1.3 pF + 0.06 pF per mm of effective length + 0.05 pF per mm of cable
Materials: Cable: Will withstand 50 kV peak for 5 minutes
Plastic Parts: UL rated 94V-0
Metal Parts: Emitters, stainless steel; bar channel, yellow chromated aluminum
Maximum ambient temperature: 75°C (167°F) maximum

Figure 5.10—Static eliminator bar

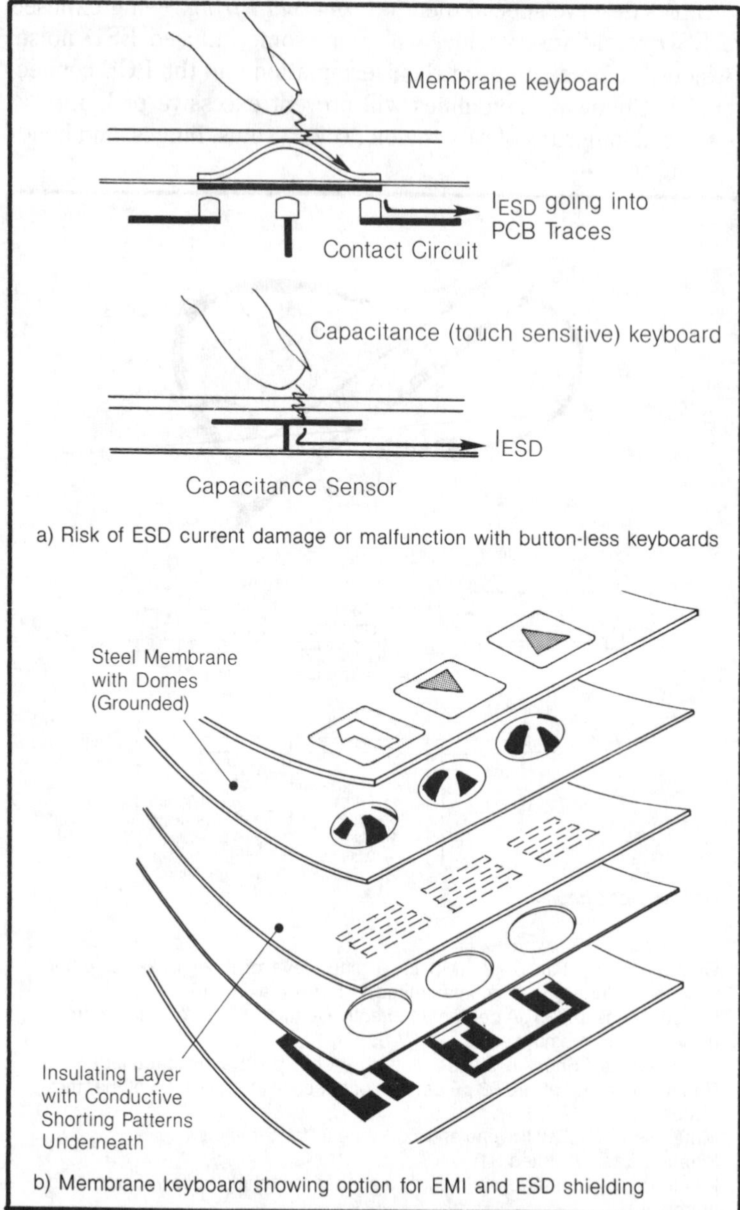

Figure 5.11—ESD problem with button-less keyboards

- do not press cable harnesses, especially flat cables, against metallic covers which are likely ESD targets. Move them away or use a thick shield
- avoid large +Vdc-to-zero volt and signal-to-zero volt loops. As for PCBs, always carry a wire close to its own return conductor.

Since, even after these precautions, an excessive ESD pick-up may still exist due to common-mode (cable-to-chassis) coupling, an interesting solution is to use CM ferrite beads and sleeves over the entire cable. This solution will work best if the circuit is a low impedance one. Add-on ferrites exist in the form of split beads and yokes. Some of them have been developed especially for flat cables (Fig. 5.12), and recent progress in lossy ferrite materials can provide for 6 or 10 dB of coupling reduction above 10 MHz, which is precisely the domain where ESD couplings become critical.

A further application of the ferrite concept is made possible by flexible ferrite tubings and wire coating (Fig. 5.12c). The compound

Courtesy of Fair-Rite Products Corporation

a) Recent ferrite beads for common-mode insertion loss

Figure 5.12a—Recent developments in ferrites

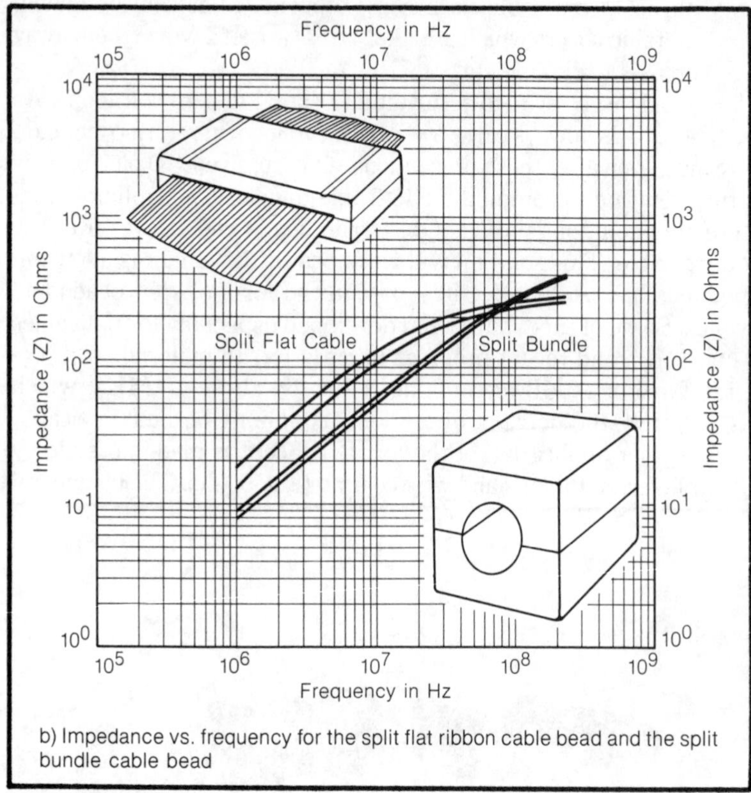

b) Impedance vs. frequency for the split flat ribbon cable bead and the split bundle cable bead

Figure 5.12b—Recent developments in ferrites

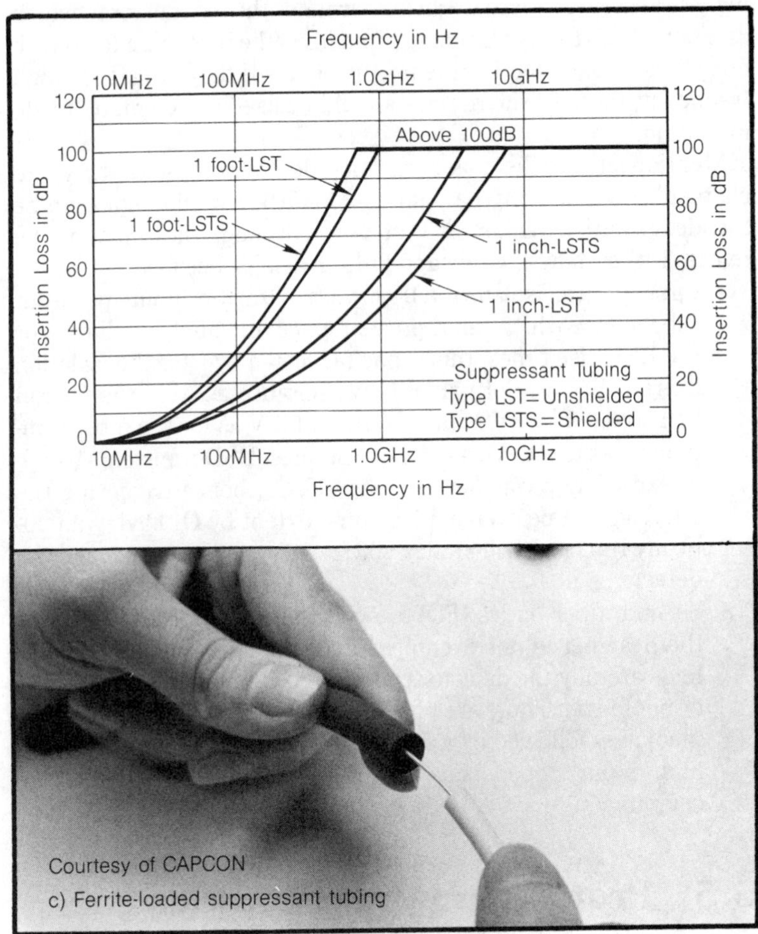

c) Ferrite-loaded suppressant tubing

Figure 5.12c—Recent developments in ferrites

makes a sort of "lossy" jacket around the conductors, which dissipates high frequency energy into heat (the best thing to do with noise!). To avoid affecting the useful signals if they are also at high frequency, these ferrite tubings should be used in a common-mode mounting.

Mention should be made of opto-electronics. Since they are claimed to be the panacea against EMI, they would appear to be the ideal solution against ESD as well. Although they must not be rejected, they may give disappointing results because:

- First, optical isolators (OI) have a parasitic input to output capacitance which can represent, once mounted, 1 to 5 pF. For the ESD rise times, this capacitance can by-pass the isolation barrier. Because of parasitic capacitance as well, the OI can be activated by dV/dt in the order of 5 V/μs for the most common parts to 500 V/μs for the best ones. ESD transients usually exceed few volts/nsec, or few kV/μs, hence triggering the OI. This can be overcome to some extent by OI having an optically transparent Faraday shield between the source and the detector.
- Second, fiber optics (FO) are certainly a better solution. But the designer is not exempted from shielding and/or decoupling carefully the detector end of the FO link. The photo-diode or photo-transistors are high impedance devices which are immediately followed by a sensitive amplifier with an extremely high input impedance, making it very sensitive to stray coupling.

5.5 Protection by the Box Design and Shielding

If the components, the PCBs, and the internal connections have been hardened to a certain ESD level V_x, and the specification requires a level $V_s > V_x$, the housing remains to make up for the difference.

The shielding precautions for the housing are basically the same as for any EMI susceptibility problem. Keep in mind, however, that the 300 MHz or higher spectrum of concern obliges the designer to consider the possible leakage of any seams and slots exceeding a few centimeters. This dictates that a machine which relies on its housing for *all its ESD hardening* should be shielded with the same precautions as any VHF equipment or sensitive radio frequency device!

In the late 1960's, when ESD problems started to become a nightmare for computers (but its mechanisms were not yet understood), some manufacturers of large computers had to take the "steamroller" approach by making all the covers RF-tight— entirely plating their frames with nickel, etc. Consequently, the mainframe took on a vault-like appearance. Besides the fact that the designer of the machine to be hardened may not be familiar with them, these techniques drastically increase the cost of the cabinet and associated hardware. They are also more likely to complicate maintenance and accessibility, and they may degrade with age.

In addition to cost, there are aesthetic/cosmetic reasons which may prohibit the use of certain shielding and gasketing techniques, particularly for consumer equipment. Typically, then, the designer will look for shielding methods which are economical, which will remain unaltered after intensive use of the equipment, and which will provide a moderate shielding effectiveness in the 10-40 dB range (i.e., a reduction of electromagnetic field by a factor of 3 to 100 times). *But the shielding method selected must provide attenuation up to the 300-1000 MHz region.*

5.5.1 Some Shielding Basics

Although full coverage of shielding theory is far beyond the scope of this book, a few guidelines are provided on how and why shields work, and examples are given of when they do not. The reader who wants to know more about the principles and applications of shields is invited to refer to the very complete textbook of Ref. 28.

Shielding effectiveness (SE) is defined as the ratio of the impinging energy to the residual energy (the part that gets through).

$$\text{For E-fields: } SE = 20 \log_{10} \frac{E_{in}}{E_{out}} \text{ dB}$$

$$\text{For H-Fields: } SE = 20 \log_{10} \frac{H_{in}}{H_{out}} \text{ dB}$$

If shields were perfect, E_{out}, H_{out} and, therefore, P_{out} would be zero. In practice, a shield is merely an attenuator which performs on two principles—absorption and reflection.

Absorption increases with:

- thickness
- conductivity
- permeability
- frequency

Reflection increases with:

- surface conductivity
- wave impedance

To evaluate absorption, or penetration losses, one needs to know how many skin depths the metal barrier represents at the frequency of concern, knowing that the field intensity will decrease by 8.7 dB (or will lose 63% of its amplitude) each time it has to go through one skin depth.

Entering all the electrical constants, we come to a simple expression for absorption loss:

$$A_{dB}=131t\sqrt{f\mu_r\sigma_r} \tag{5.1}$$

where

t = thickness of conductive barrier in mm
f = frequency in MHz
μ_r = permeability *relative* to copper
σ_r = conductivity (the inverse of resistivity) relative to copper.

For instance, a 0.03 mm (1 mil) aluminum layer will offer at an absorption loss at 100 MHz of:

$$A_{dB}=131\times0.03\sqrt{100\times1\times0.6}\cong30.4\text{ dB}$$

This is equivalent to a reduction of the field strength by a factor of $(10)^{30.4/20}\cong33$ times.

Looking at Eq. 5.1 leads to a few remarks:

a) *for non-magnetic materials* (μ_r=1), the penetration losses increase with conductivity σ_r. Since no metal has better conductivity than copper (except for silver with σ_r=1.05), any non-magnetic metal will show less absorption than copper. Zinc, for instance, for which

σ_r=0.3, will exhibit for a thickness of 0.03 mm (1 mil) an absorption loss at 100 MHz of:

$$A_{dB}=131\times 0.03\sqrt{100\times 1\times 0.3}\cong 20\text{ dB}$$

b) *for magnetic materials* (μ_r>1), the penetration losses increase with μ_r. On the other hand, their conductivity is less than copper's. Since μ_r for steel or iron is in the range of 300 to 1000, while σ_r is about 0.17, a definite advantage exists for magnetic materials. However, above a few hundred kHz (ferrites excepted), μ_r generally collapses to equal 1, while σ_r is still mediocre.

To evaluate reflection, it is necessary to know if the shield is in near or far-field conditions. Near-field conditions, where the shield is closer than λ/6 to the source, are the most critical. For pure electric fields, since their wave impedance is high, it is relatively easy to get good reflection properties because the field-to-shield mismatch is large. For near magnetic fields, the wave impedance is low and it is more difficult to get good reflection.

For near-field conditions, the reflection losses for E-Fields are equal to:

$$R\text{ dB}_{(e)}=20\log_{10}\left[\underbrace{\frac{120\pi}{4Z_b}}_{\text{far-field reflection term}}\times\underbrace{\frac{\lambda}{2\pi r}}_{\text{near-to-far-field correction}}\right] \tag{5.2}$$

$$=20\log_{10}\frac{4500}{rFZ_b} \tag{5.2a}$$

For H-Fields, reflection losses are equal to:

$$R\text{ dB}_{(H)}=20\log_{10}\left[\underbrace{\frac{120\pi}{4Z_b}}_{\text{far-field reflection term}}\times\underbrace{\frac{2\pi r}{\lambda}}_{\text{near-to-far-field correction}}\right] \tag{5.3}$$

$$\cong\frac{2rF}{Z_b} \tag{5.3a}$$

where

Z_b = Shield barrier impedance in ohms/square
F = frequency in MHz
r = distance from radiating source in meters

How does one know if, at distances $<<\lambda$, the field is more electric or magnetic in nature? By looking at the radiating source, one might have an idea of the predominant mode: Circuits switching large currents, such as power supplies, solenoid drivers, and heavy current logics, generate strong magnetic fields. Conversely, voltage driven high-impedance or open-ended lines create electric fields.

If we apply the specific conditions of ESD to the properties of shields, we see that:

- The most threatening part of the ESD spectrum being in the high frequency region, practically any barrier having a thickness of 0.1 millimeter or more and made of a homogeneous metal will provide excellent absorption loss.
- Conductive paints or coatings will not perform as well, by absorption, because their film thickness represents only a few skin depths or less. Therefore, their absorption loss is rather low. Here, again, conductivity makes the difference. Silver, copper and zinc films will still provide absorption loss in the range needed for ESD; i.e., 10 to 40 dB. Graphite paints will provide marginal or null absorption loss because of their relative conductivity of 10^{-5} to 10^{-6}.

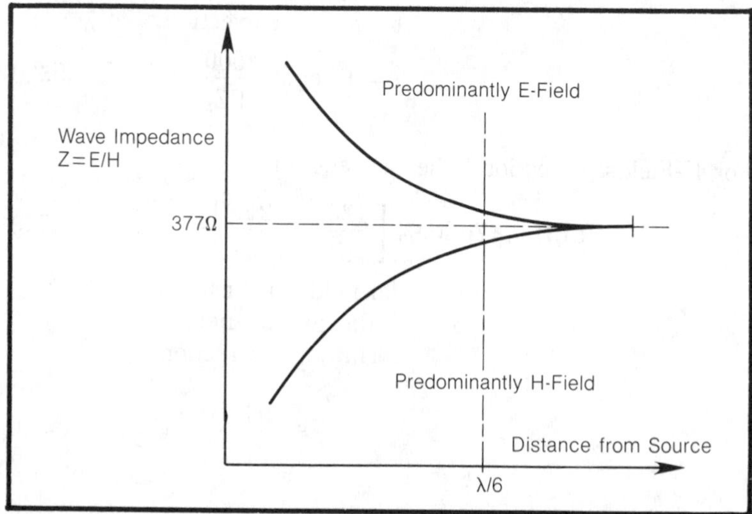

Figure 5.13—Conceptual illustration of near vs. far fields

- As far as reflection, the ESD situation complicates the issue: personnel ESD, coming from a source whose total impedance is larger than 377 ohms, will tend to generate high impedance fields (i.e., predominantly E Term) while furniture ESD will generate low impedance fields (i.e., predominantly H Term). Now, how near is this near-field? Since the "nearness" is expressed as a ratio of $\lambda/2\pi r$, we have a frequency dependent term; that is, the lower portions of the spectrum will be in an extreme near field condition, while the upper portions where λ is shorter will approach or even pass the near-far transition condition of

$$r = \frac{\lambda}{2\pi}$$

This is discussed in Sec. 2.4.1 with the experimentation on ESD radiation. Two situations may occur:

1) The discharge occurs not on the conductive housing of the equipment itself, but rather, it occurs nearby. In this case, the discussion on absorption and reflection applies. Furniture discharges are the most threatening because this is where the magnetic field predominates and the reflection term will be minimal.
2) The discharge occurs right on the housing (direct ESD). In this case the previous approach on near-field and reflection losses is more questionable. What is impinging the metal barrier is not a field but is already a current. This current then penetrates the metal barrier, is attenuated, and what is left on the other side re-radiates as shown in Fig 5.14. Therefore, the strict definition of shielding effectiveness as a ratio of two fields does not apply here. A better measure would be to use the transfer impedance of the shield, which would convert the current on the discharge side to a voltage on the inner side (Ref. 29). This voltage in turn excites a radiation mechanism inside the housing. This mechanism can also be expressed as the ratio of the re-radiated field to the ESD current, i.e.:

$$\frac{E \text{ volts/m}}{I \text{ amp}}$$

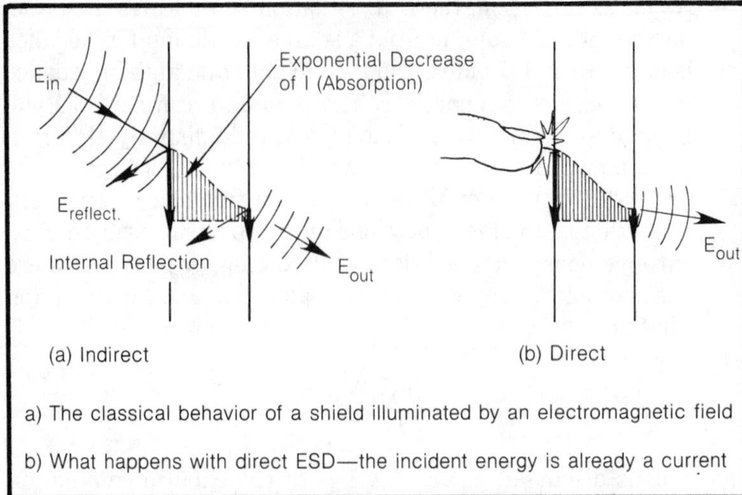

Figure 5.14—Model of shield barrier against ESD

This ratio would have the dimension of a radiation impedance in ohms/meters.

Finally, another way to consider shielding effectiveness against ESD is shown in Fig. 5.15. In this case, an ESD event is created by substituting a straight conductor to the shield, measuring the effects on the victim, then putting in the actual shield and comparing the results. This method has the advantage of showing that with a conductive metal barrier, a reflection actually takes place.

In any case, this direct ESD mechanism suggests that, with low impedance sources like furniture, the first reflection term is rather small and the only chance a shield has of performing well is to have either excellent conductivity, as close as possible to copper, or sufficient thickness to represent at least a few skin depths. Table 5.1 shows the absorption loss of several metals at several thicknesses for the 100 MHz region.

Housings, however, are not made like continuous metal cubes. They have slots, seams, apertures, etc. which will inevitably leak.

As in a chain, a shield is only as good as its weakest link; therefore, it is important to know the weak points in the shields in order to establish some realistic objectives.

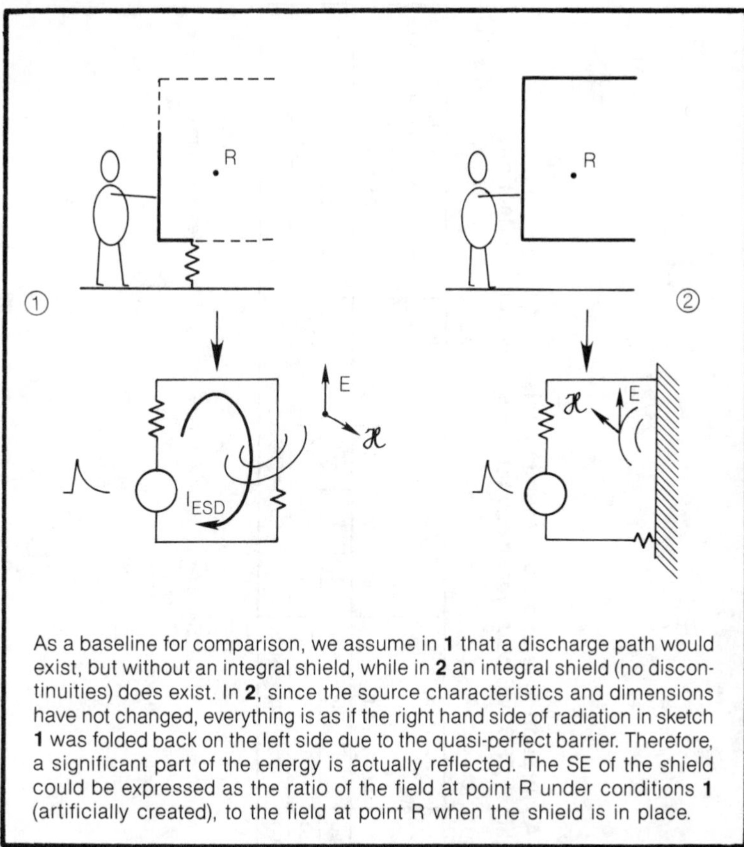

As a baseline for comparison, we assume in **1** that a discharge path would exist, but without an integral shield, while in **2** an integral shield (no discontinuities) does exist. In **2**, since the source characteristics and dimensions have not changed, everything is as if the right hand side of radiation in sketch **1** was folded back on the left side due to the quasi-perfect barrier. Therefore, a significant part of the energy is actually reflected. The SE of the shield could be expressed as the ratio of the field at point R under conditions **1** (artificially created), to the field at point R when the shield is in place.

Figure 5.15—Conceptual view of shield reflection in the case of direct ESD

- At low frequencies, what counts is the nature of the metal used—its thickness, conductivity, and permeability.
- At high frequencies, where any metal would provide hundreds of dB of shielding, they are never seen because seams and discontinuities completely spoil the metal barrier.

A slot in a shield can be compared to a slot antenna which, except for a 90° rotation, behaves like a dipole (see Fig. 5.16).

Figure 5.17 shows the leakage caused by long seams or slots, assuming worst-case polarization. It is assumed that the slot has practically no depth (i.e., the thickness of metal is smaller than the

Table 5.1—Absorption loss in dB of several metals.
The shielding effectiveness against ESD will be *at least* equal to this plus some reflection.

Thickness (mm)	Copper			Aluminum			Zinc			Steel*			Nickel* σ_r=0.2		Copper Paint (non-homogeneous metal) σ_r=.04
	0.01	0.1	1	.01	0.1	1	.01	0.1	1	.01	0.1	1	.01	1	.05 mm (2 mil)
30 MHz	7	70	700	5.2	52	520	4	40	400	3	28	280	3	31	7
100 MHz	13	130	>1000	9.5	95	950	7	72	720	5	50	500	6	58	13
300 MHz	22	220	>1000	17	170	>1000	12	125	>1000	9	88	880	10	98	22

*Although steel and nickel are magnetic materials, their relative permeability μ_r collapses to ≅1 above a few MHz. This has been accounted for.

**Values above 150 dB are calculation results since they obviously could not be measured.

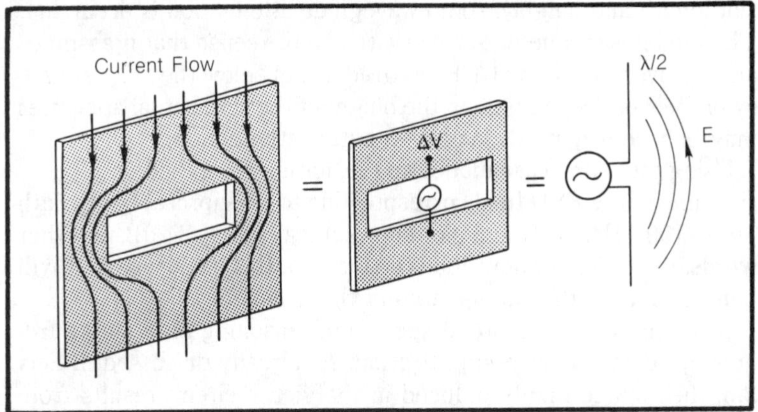

Figure 5.16—Radiation caused by a discontinuity in the shield

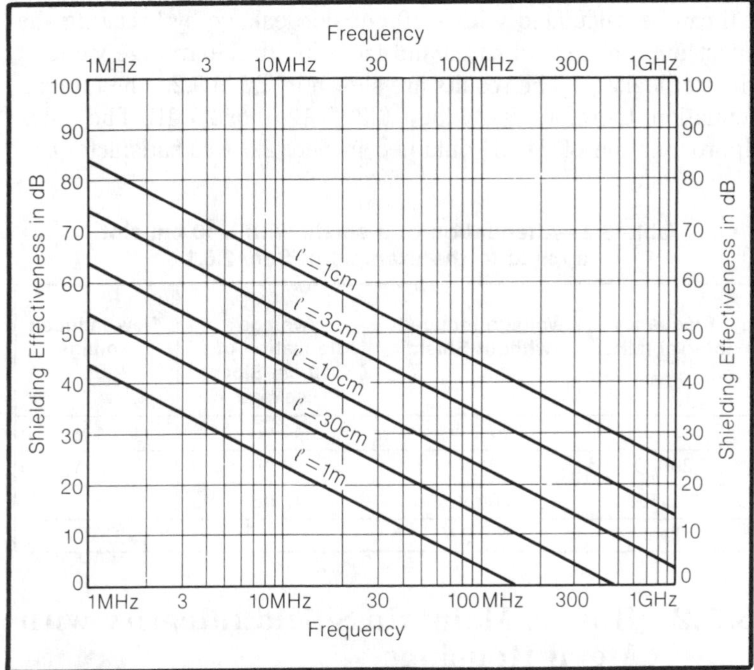

Figure 5.17—Shielding effectiveness corresponding to slot leakage for different slot lengths

slot length and height), so no waveguide attenuation is occurring. The model is simple but conservative in the sense that it assumes an improvement ratio of 1/F (20 dB/decade) below the λ/2 resonancy of the slot. Depending on the height of the slot, actual apertures may behave differently and give better attenuation.

Using this worst-case model we see, for instance, that for a 3 nsec rise time of the ESD field corresponding to a reciprocal bandwidth $1/\pi\tau_r$=100 MHz, a 10 cm slot shows a leakage of 24 dB. In other words, regardless of how good the metal of this enclosure is, it will behave as a 24 dB barrier around this frequency.

Since the ESD is a broad spectrum, reasoning on a single frequency seems an oversimplification. As already discussed in Sec. 2.6, the voltage finally induced in the victim circuit results from a cascade of mechanisms with frequency dependent slopes, and the slot leakage becomes one of them. However, the example of Sec. 2.6 can be calculated with a 10 cm slot leakage by breaking the spectrum into several slices and factoring in the average value of the slot leakage. The results are shown in Table 5.2. The total attenuation, therefore, is 20 $\log_{10}$ (222.5/12.6) or 25 dB. The quick approximation of 24 dB obtained in Sec. 2.6 was satisfactory.

Table 5.2—Attenuation of a shield with a 10 cm slot, applied to the example of Sec. 2.6.1

Frequency Interval (MHz)	Voltage Induced without Shield	Average Attenuation of a 10 cm Slot Leakage	New Induced Voltage
3-30	67.5 V	44 dB	.4 V
30-100	91 V	29 dB	3.2 V
100-500	64 V	17 dB	9 V
Total	222.5 V		12.6 V

5.5.2 How to Maintain Shield Integrity with Metal Housings

A metal housing already has the basic advantage of being a naturally efficient barrier. All the talent of the designer should be aimed at not spoiling this barrier with excessive leakages:

- *All* metal parts should be bonded together. A floated item is a candidate for re-radiation.
- For cover seams, slots, etc., how frequently they should be bonded is a matter of design objective. Figure 5.17 shows that a 10 cm leakage is worth about 20 dB of shielding. If the goal is closer to 30 or 40 dB, seams or slots should be broken down to 3 cm or 1 cm. For permanent or semi-permanent closures, this means frequent screws or welding points or an EMC conductive gasket. For covers, hatches, etc., this means flexible contacts or gaskets.

The following is an organization of these solutions in an orderly sequence. As efficiency increases, cost increases as well.

- If only minimal shielding effectiveness is needed (the 0-20 dB range), the simplest technique is to have frequent bonding points and, for covers, short flexible straps made of flat braid or copper foil as shown in Fig. 5.18. This solution does bond only on the hinge side, but if no sensitive items are located near

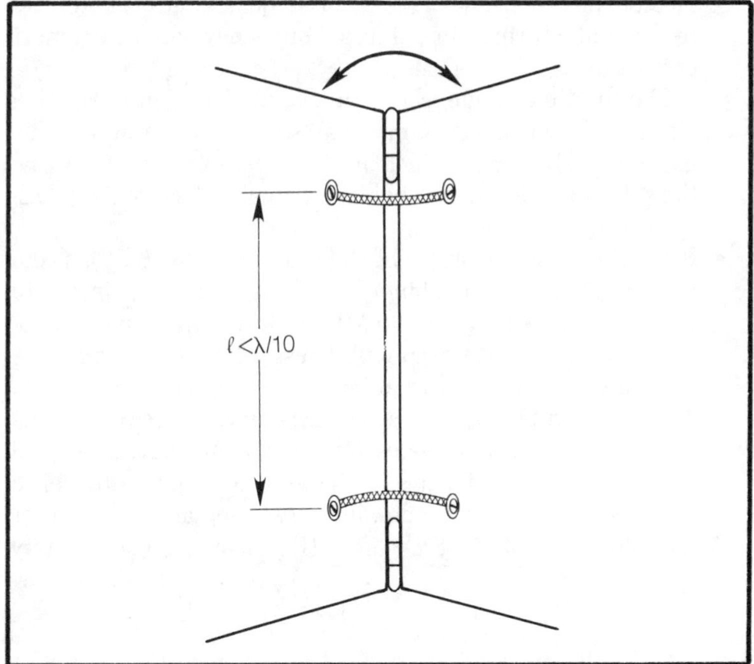

Figure 5.18—Braided jumpers screwed on paint-free areas

the opposite side of the seam, this can be sufficient. For the opposite side, a wise precaution is to use a grounded lock or fastener.

The λ/10 shown in Fig. 5.18 means that, for a minimum ESD rise time of 1 nanosec, the distance between jumpers should stay within 5 cm for a 20 dB shielding objective, and up to 30 cm if 6 dB is sufficient.

- If bonding at the hinged side only leaves an excessive length of ungasketed seams, additional bonding points are necessary. In this case, the techniques shown in Fig. 5.19 can be used.

  Figure 5.19a shows a few soft springs scattered along the edges of the cover. For durable performance, the spring contact riveting *must be corrosion-free,* which may render that solution more difficult to apply than it would seem.

  A variation of this, shown in Fig. 5.19b, is to use sections of spring contacts called finger-stocks. Several types of finger-stocks are available, such as low-pressure, knife edge, and medium pressure. They require an adequate control of pressure by manufacturing tolerances, but they are extremely dependable.

  The third technique, shown in Fig. 5.19c, is an interesting alternative which takes minimal surface preparation. The grounding "buttons," which are fairly compliant to gap variations due to their spring loading, are mounted simply by press-fit or threaded stud.

- If a higher grade of shielding is required (20 to 40 dB), a continuous conductive bonding of seams is necessary, since a 40 dB maximum leakage at 300 MHz (λ/2=50 cm) would require screws or rivets every 0.5 cm! These continuous conductive joints are available in several forms and stiffnesses (Fig. 5.20). The hollow rubber gasket is less expensive to use because its wide elasticity compensates for large joint unevenness and warpage. The price for this is a lesser contact pressure, hence higher resistivity. Therefore, it is best used as a solution for the lower side of the SE range. Here again, a good quality

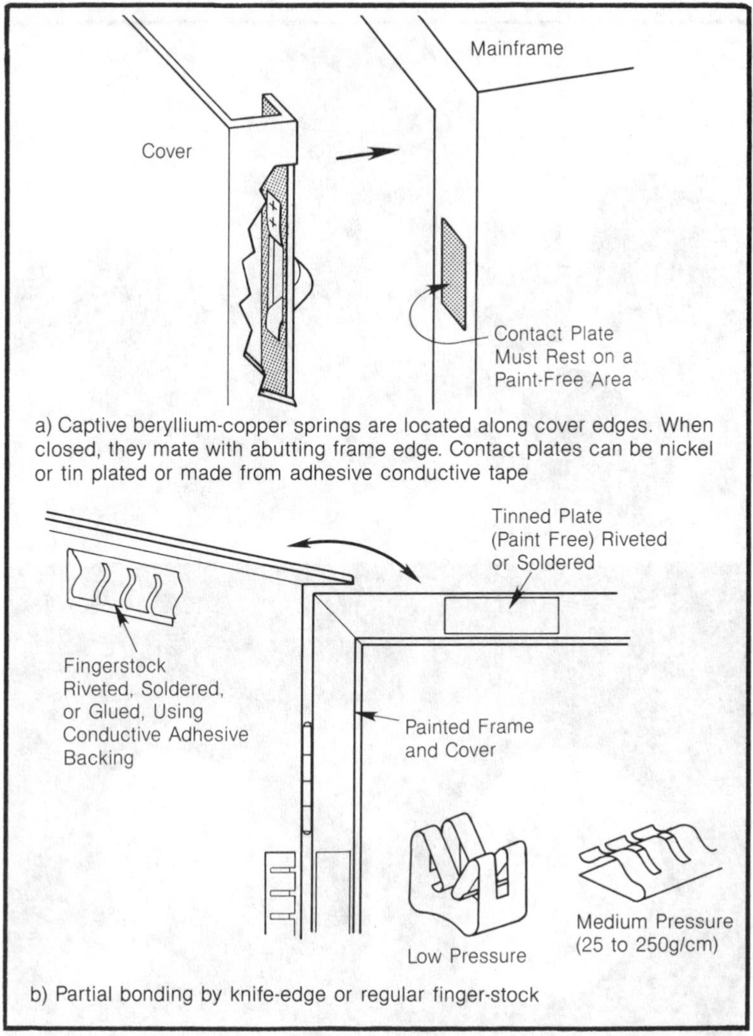

Figure 5.19a—Solutions to maintain shield continuty with painted metal box

Courtesy of Chomerics

c) Grounding buttons allow multiple grounding with a contact resistance within 2-20 mΩ, and they do not require special preparation of mating frames

Figure 5.19b—Solutions to maintain shield continuity with painted metal box

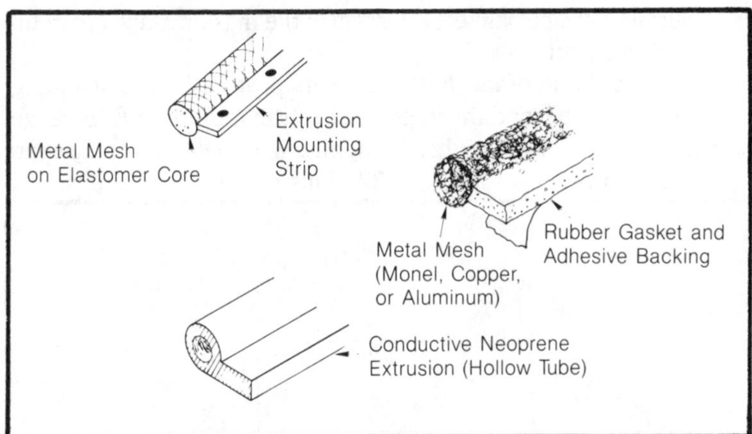

Figure 5.20—Compressible conductive gaskets

mating surface can be made by applying conductive tape (Fig. 5.21) over the metal surface *before painting.* Then a piece of masking tape is pressed over the conductive foil and the metal

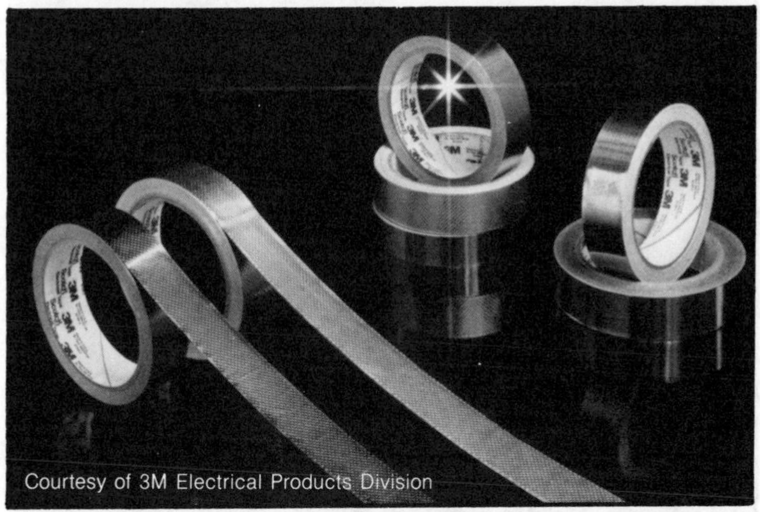

Figure 5.21—Conductive tapes used to create a good conductive area for local shielding or contact point. Surface resistance can be as low as a few mΩ/square.

surfaces can be painted, after which the masking tape is carefully peeled off.

Metal braid of mesh type gaskets provide higher shielding, close to or beyond the upper side of the required SEdB range.

- Finally, if an even higher hardening is necessary, the ultimate solution is shown in Fig. 5.22. This solution is the most effi-

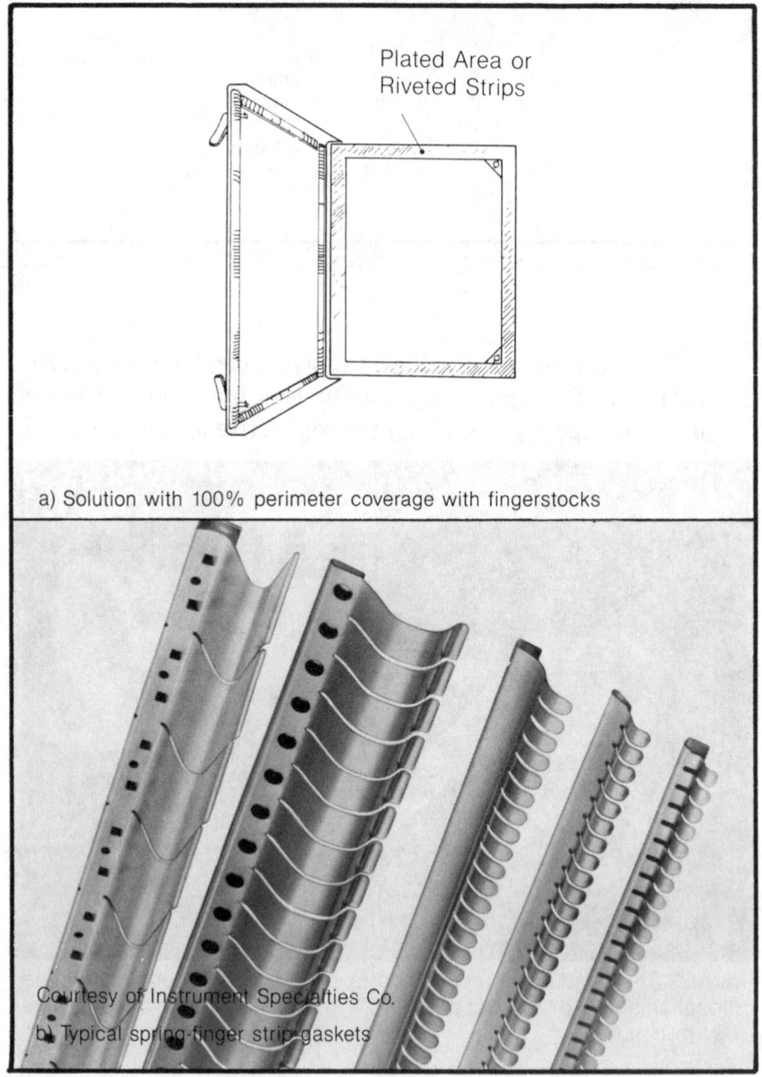

a) Solution with 100% perimeter coverage with fingerstocks

b) Typical spring-finger strip gaskets

Figure 5.22—The ultimate solution for hardening greater than 40 dB

cient since 100% of the seam becomes a very good conductive joint. Besides its cost, it adds the need for a strong locking mechanism to ensure good, even pressure on all of the spring blades. This method is applicable to both rotating (hinged) or slide-mating surfaces.

Independently of bonding, a reduction of the slot leakage can be obtained, at practically no cost, by designing the cover edges so that they always overlap, as in Fig. 5.23. This overlap will act somewhat as a waveguide-beyond-cutoff; i.e., it will provide some attenuation.

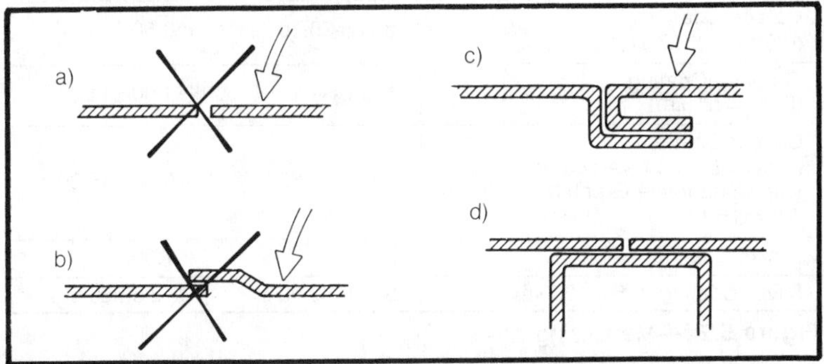

Figure 5.23—Design of cover edges for improved ESD immunity

5.5.3 How to Get Shield Integrity with Plastic Housings

Plastic housings provide no shielding whatsoever. Therefore, unless the PCBs and internal wiring have been hardened sufficiently, the plastic must be made conductive. Several metallizing processes exist which are summarized in Fig. 5.24 along with their average 1985 cost. Since, as discussed in Sec. 5.5.1, conductive coatings exhibit a rather mediocre absorption loss, their only chance to work is by reflection. Based on reflection loss only, Fig. 5.25 shows the shielding effectiveness of thin coatings versus their distance/wavelength relation with the source. (A more detailed explanation can be found in Ref. 28.) If a shielding effectiveness in the range of 30-40 dB is desired, especially against low impedance sources like furniture ESD, a conductive process with 1 Ω/square or less must be selected. With a close look at the box design, some features of plastic housing can be turned into an advantage over

Process	Cost (1985) Includes application cost (for paints/spray, manual operation assumed) Dollars/sq. ft.	(Dollars/sq. m)
Electroplating	\$0.25 - \$2.00	(\$2.5-\$20.00)
Electroless Plating (Copper and Nickel, Inside and Outside	\$1.30	(\$13.00)
Metal Spray	\$1.50-\$4.00	(\$15.00-\$40.00)
Vacuum Deposition	\$0.50-\$2.25	(\$5.00-\$22.00)
Carbon Coating (0.5 mil = 12.7 μm)	\$0.05-\$0.50	(\$0.50-\$5.00)
Graphite Coating (2 mil = 51 μm)	\$0.10-\$1.10	(\$1.00-\$11.00)
Copper Coating (2 mil = 51μm) Overcoated with Conductive Graphite at 0.2 mil = 5μm	\$0.30-\$1.50	(\$3.00-\$15.00)
Nickel Coating	\$0.70	(\$7.00)
Silver Coating (0.5 = 12.7μm)	\$1.25-\$3.00	(\$12.5-\$30.00)

Figure 5.24—Metallizing costs

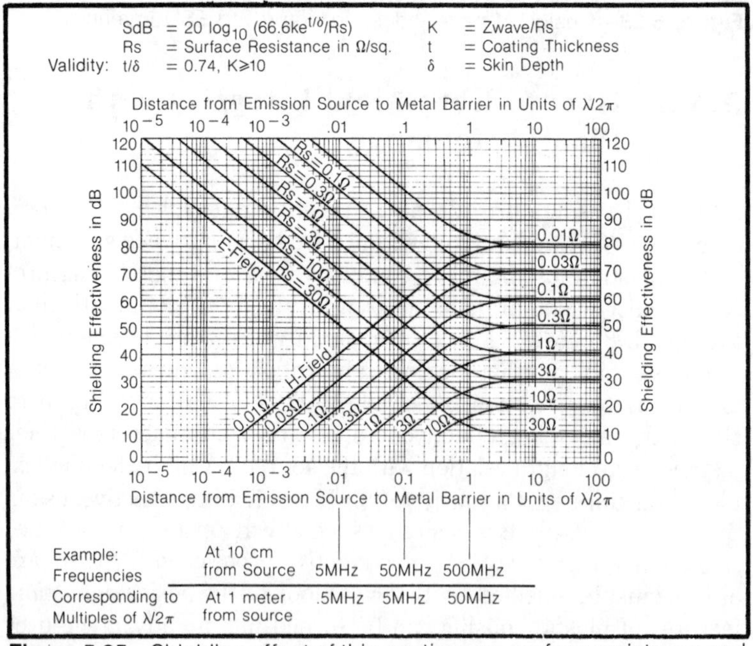

Figure 5.25—Shielding effect of thin coatings vs. surface resistance and near/far-field conditions

ESD, and some pitfalls can be avoided which are summarized in Fig. 5.26.

Besides these particular aspects, all that has been said for metal housings (slots, seams, bonding, etc.) applies to metallized plastic

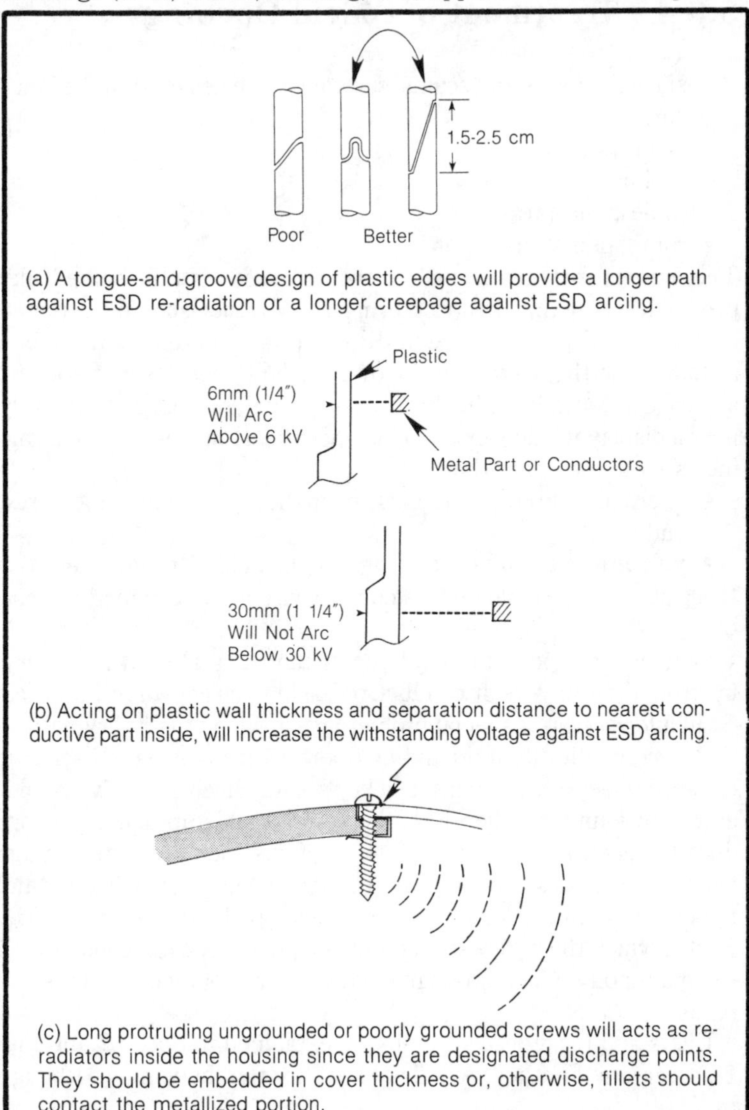

1.5-2.5 cm
Poor
Better
(a) A tongue-and-groove design of plastic edges will provide a longer path against ESD re-radiation or a longer creepage against ESD arcing.
Plastic
6mm (1/4″)
Will Arc
Above 6 kV
Metal Part or Conductors
30mm (1 1/4″)
Will Not Arc
Below 30 kV
(b) Acting on plastic wall thickness and separation distance to nearest conductive part inside, will increase the withstanding voltage against ESD arcing.
(c) Long protruding ungrounded or poorly grounded screws will acts as re-radiators inside the housing since they are designated discharge points. They should be embedded in cover thickness or, otherwise, fillets should contact the metallized portion.

Figure 5.26—Plastic covers design for better ESD immunity

as well. Here, too, surface conductivity can be improved locally by using copper or aluminum tapes.

5.5.4 Treatment of Shield Openings

Besides the joints and seams, several large holes exist in the housings for:

- displays
- cooling
- cable penetrations
- components shafts, etc.

These apertures create shield discontinuities with the same leaky properties as seams. In strict terms of EMI, they should be treated to provide SE performance in dB equal to or better than the one required for the whole box. However, ESD bears some specific aspects. It is rather unlikely that people will discharge frequently near a display window or a cooling aperture. However, if the packaging is so unfortunate that:

- likely discharge points exist near these functional apertures, and
- vulnerable circuits are located right behind them, then

these apertures should be treated with wire mesh, conductive glass, etc.

Switches and potentiometer shafts can be treated, if necessary, by grounding fingers or conductive bushing as shown in Fig. 5.27. Switch locks pose a special problem because, by their very nature, they receive the tip of a hand-held key, which acts as a discharge enhancer (see Sec. 1.2.2) inside a key socket which is poorly grounded or ungrounded. Therefore, the key slot ends up making an arc duct to reach the electrical portion of the lock. To make things worse, the lock is probably the first thing the user will zap onto. Special anti-static locks have been developed like the one in Fig. 5.28 in which the rotary socket makes a good electrical contact with the main body which, in turn, makes a large contact area with the panel.

LEDs and incandescent bulbs, as explained in Sec. 2.2 and Fig. 2.7b, create another Achilles' heel in the housing. Although

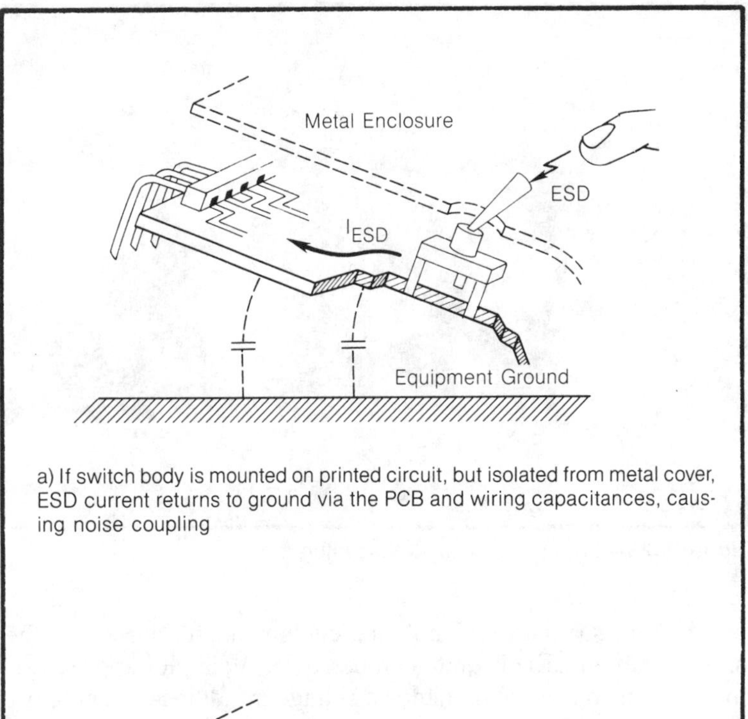

a) If switch body is mounted on printed circuit, but isolated from metal cover, ESD current returns to ground via the PCB and wiring capacitances, causing noise coupling

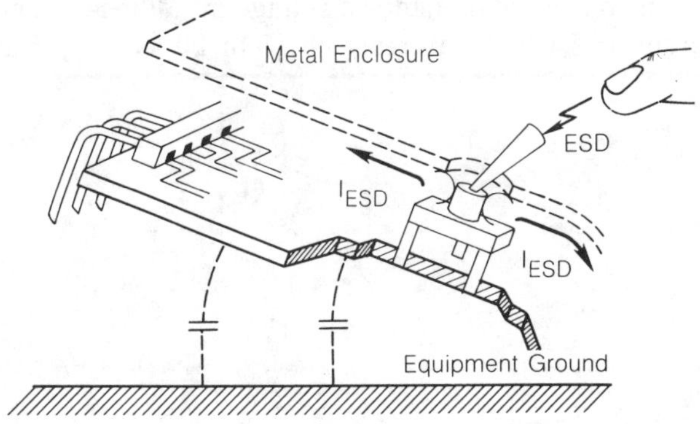

b) If switch body is connected to metal cover, ESD current flows on metal surface (another solution is to use plastic toggles or metal bracket ground pushbuttons separately from keyboard)

Figure 5.27—ESD penetration via switches and metallic shafts

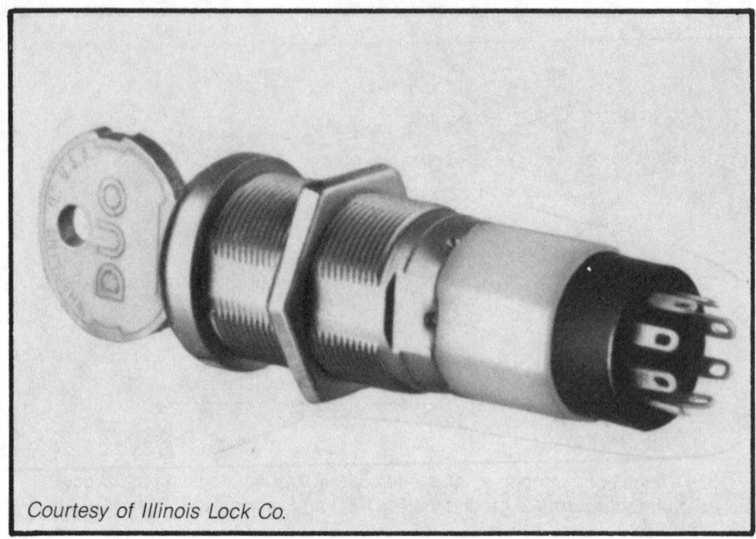

Courtesy of Illinois Lock Co.

Figure 5.28—Example of anti-static switch lock

feasible—in sophisticated military equipment, for instance—the shielding of an LED is quite cumbersome. A simpler approach is to merely increase the breakdown voltage by adding an insulating transparent lens which can resist to 15 or 20 kV (Fig. 5.29).

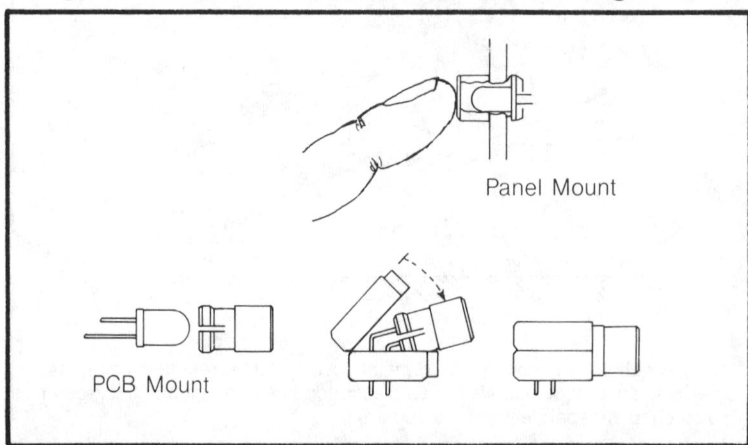

Figure 5.29—LED protection by an ESD protective lens. This CLIPLITE® from Visual Communication Co. will withstand 16 kV.

Cable penetration will be discussed in Sec. 5.6 since the cable shield termination at the housing entry is a key factor in ESD immunity.

Summary of ESD Protection Measures at Packaging/Box Level

- For metal housings:
  - —bond together *every* metal part (a floated item is a candidate for re-radiation)
  - —avoid long seams and slots: a 10 cm seam is totally leaky at upper frequencies of ESD, or
  - —use gasket and waveguide effect
- For plastic housings,
  - —use conductive coating, ≤1 Ω/square, then treat like a metal housing
  - —avoid long protruding screws inside
  - —try to take advantage of air-gap barrier and thickness factor (works for you, costs nothing)
- Respect shield integrity at cable entry points
- Power-mains protection
  - —use surge suppressors
  - —filter mains with CM/DM filters
  - —use absorptive cable for mains cord
  - —use shielded power cord, bonded to frame
- Parts location
  - —locate sensitive wiring inboard
  - —locate less sensitive components outboard

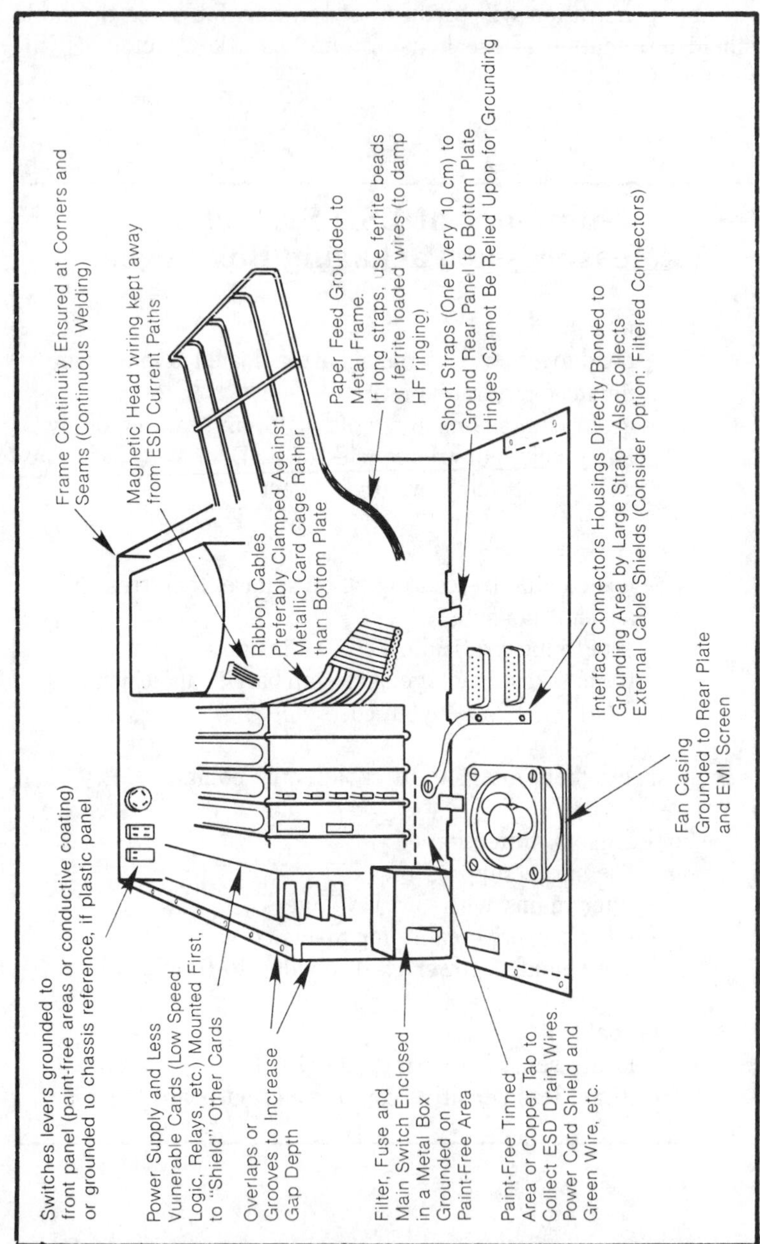

Figure 5.30—A menu of ESD protection measures at packaging/shielding level, illustrated by a fictitious machine

5.6 ESD Protection of External Cables

External cabling poses a much greater problem. Due to their direct illumination during ESD, external cables become unintentional antennas and convert the radiated field into induced voltages and currents. Fig. 5.31 summarizes what happens to external cables during an ESD event.

Among the external cables, flat cables top the list of potential ESD carriers because they are generally untwisted, unshielded, and terminated on plastic connectors right on PCB inputs. They also offer a larger, more even stray capacitance to ground than multipair cables, for instance, where wire pairs are convoluting randomly into the harness.

In a well documented set of experiments, Charlotte Palmgren (Ref. 30) has measured the voltage induced in flat cables by ESD applied to the cabinet where these cables were connected.*

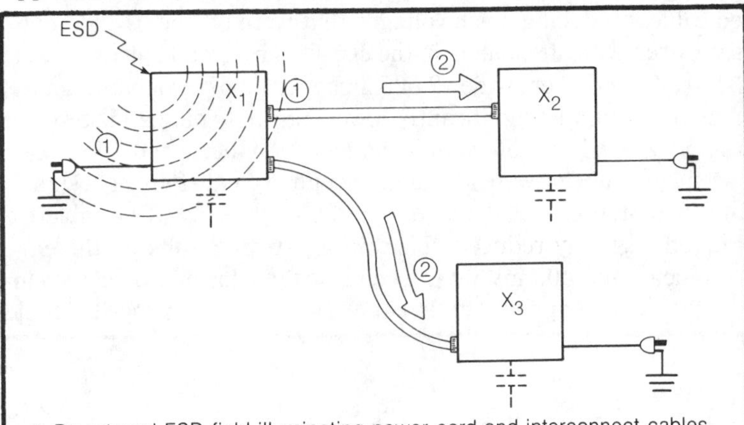

1. By external ESD field illuminating power cord and interconnect cables (field-to-cable, CM and DM)
2. By direct ESD currents sharing interconnect cables as alternate return paths (common-mode current if exposed wires, or transfer impedance coupled voltages if shielded cables)

If units $X_{1\cdot 2\cdot 3\cdots n}$ are from different types or different manufacturers, the system will be *as weak as the lowest ESD fail level.*

Figure 5.31—Contribution of external cables to ESD coupling

*Based on the paper "Shielded Flat Cables for EMI and ESD Reduction" by C. Palmgren appearing in 1981 IEEE International Symposium on Electromagnetic Compatibility, August 18-20, 1981, Boulder, Colorado, pp. 281-287. ©1981 IEEE.

In her ESD test, Palmgren simulated a peripheral cabinet connected by an I/O cable to a control cabinet or console. The cable's sensing line in the cable was terminated in its characteristic impedance at the peripheral end. The noise pulse at the sensing end was displayed on a Tektronix 7834 storage oscilloscope with a 50 ohm input impedance. The oscilloscope was shielded by placing it inside the console. The peripheral cabinet and the console were both grounded to the electrical systems ground with green wires. This test set-up is shown in Fig. 5.32. The discharge capacitor, C_D, is charged by a 10 kV source from 110 MΩ. Capacitor C_D is then discharged through resistor R_D to the peripheral cabinet through an adjustable ball air gap set to break down one or two times per second (at least 22 charging time constants in all cases). The oscilloscope records the noise voltage induced in the signal conductors by the discharge.

The standard 3M ESD test with 400 pF and 100 Ω was used to compare four families of ribbon cable. As expected, the electromagnetic field of the discharge coupled directly into the unshielded cable, producing noise voltages that were off the oscilloscope's scale even with attenuators in the circuit. The ground plane of cable 3460 (50 conductors wide=2.625 inches or 6.667 cm) reduced noise pickup enough to significantly lower induced voltage pickup, but the noise levels of 150 volts on an edge conductor and 40 volts on a center conductor were still high enough to create false data signals and potential circuit damage. The 360° shielded 3517 cable exhibited a greatly reduced noise voltage with 3 volts on the edge conductor and 400 mV on the center conductor of a 50 conductor cable. The experimental cable with an improved shield

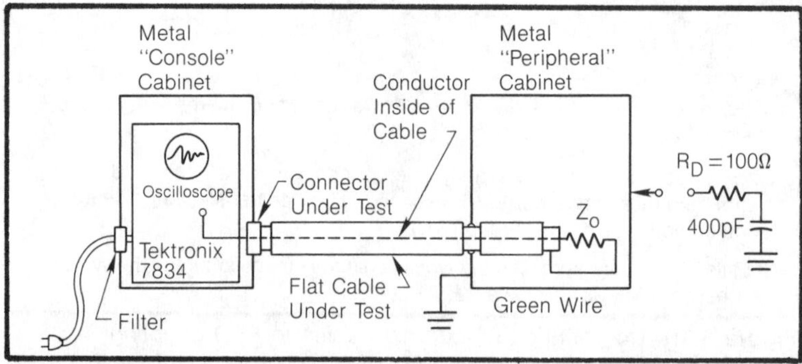

Figure 5.32—Test set-up used by C. Palmgren (Ref. 30) to measure ESD induced noise into flat cables

demonstrated noise voltage pickups of only 400 mV on an edge conductor and 200 mV on the center conductor of a 60 conductor cable (Fig. 5.33).

The shielded and ground-plane cables showed a difference in noise voltage levels on edge and central conductors. Figure 5.34 gives an edge-to-center profile of noise levels on 3517 cable. For a 360° wrapped shield, the voltage pickup started at 3.0 volts then rapidly dropped off to its minimum of 400 mV by the seventh conductor from the edge. This observed edge-to-center difference can be attributed to the sharpness of the bend in the cable shield where it is wrapped about the edge of the cable. The electromagnetic field

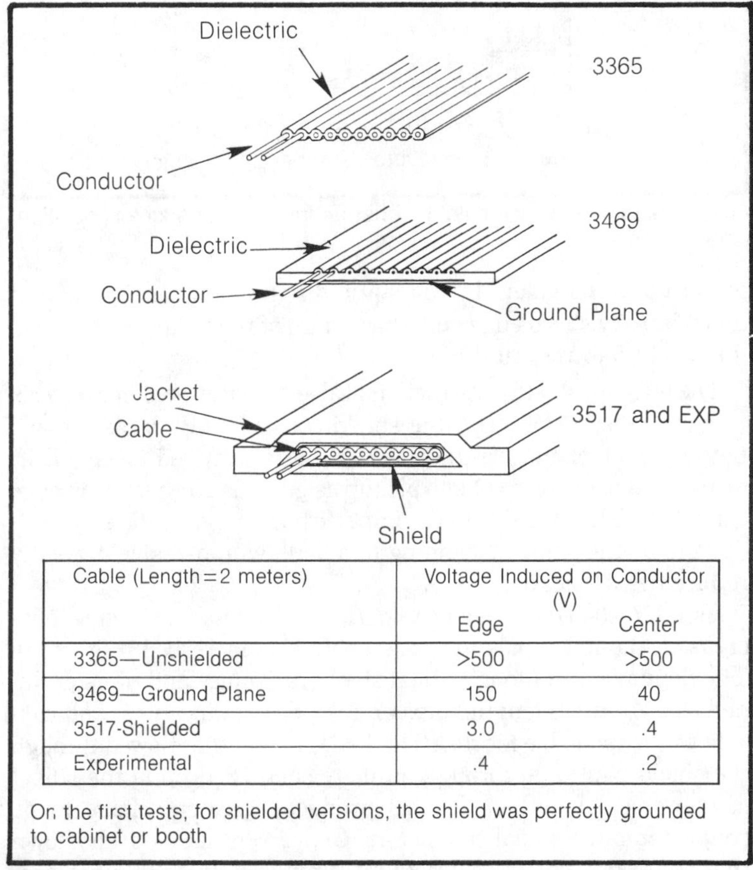

Cable (Length=2 meters)	Voltage Induced on Conductor (V)	
	Edge	Center
3365—Unshielded	>500	>500
3469—Ground Plane	150	40
3517-Shielded	3.0	.4
Experimental	.4	.2

On the first tests, for shielded versions, the shield was perfectly grounded to cabinet or booth

Figure 5.33—ESD noise pickup on 2 meters of several cable types, for 10 kV ESD

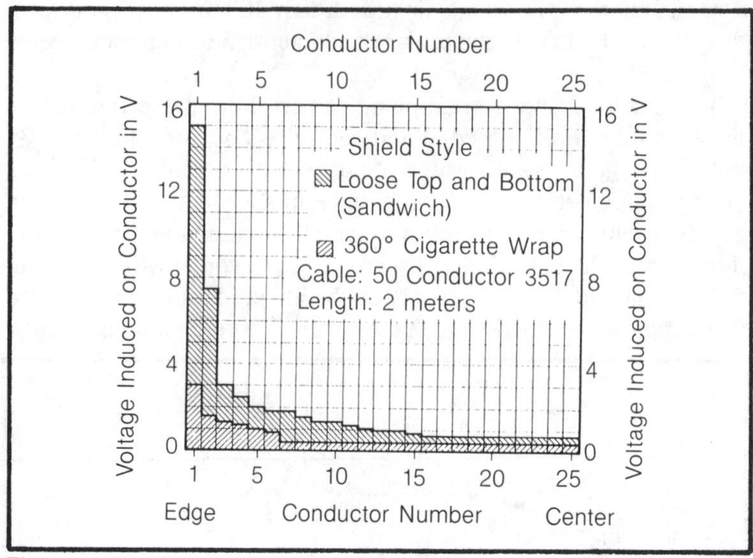

Figure 5.34—Effect of shield construction and conductor position, edge-to-center

produced in the shield by the static discharge is concentrated at this relatively sharp edge and, thus, more of the field can penetrate the shield into the cable.

The effect of shield construction is also demonstrated in Fig. 5.33. In early versions of 3517, the shield consisted of a loose top and bottom layer of mesh rather than a 360° wrapped shield. ESD testing revealed noise voltage pickup as great as 15 volts on an edge conductor. This was due to the imperfect coverage of the edge conductors by the shield. Changing to a 360° wrapped shield greatly reduced ESD pickup.

As the distance between the peripheral cabinet and console increased, the noise pickup in the cable also increased (Fig. 5.35). The dependence of longitudinal shield resistance and transfer impedance Z_t on the length between the peripheral cabinet and console was responsible for this. The ESD current that flowed through the shield had to go through an increasing IR drop in the shield as the shield lengthened. This produced greater voltage to be coupled into the signal conductors. Conversely, it has been shown that increasing shield conductivity through the use of more conductive materials decreased a cable's ESD sensitivity. *Even in two*

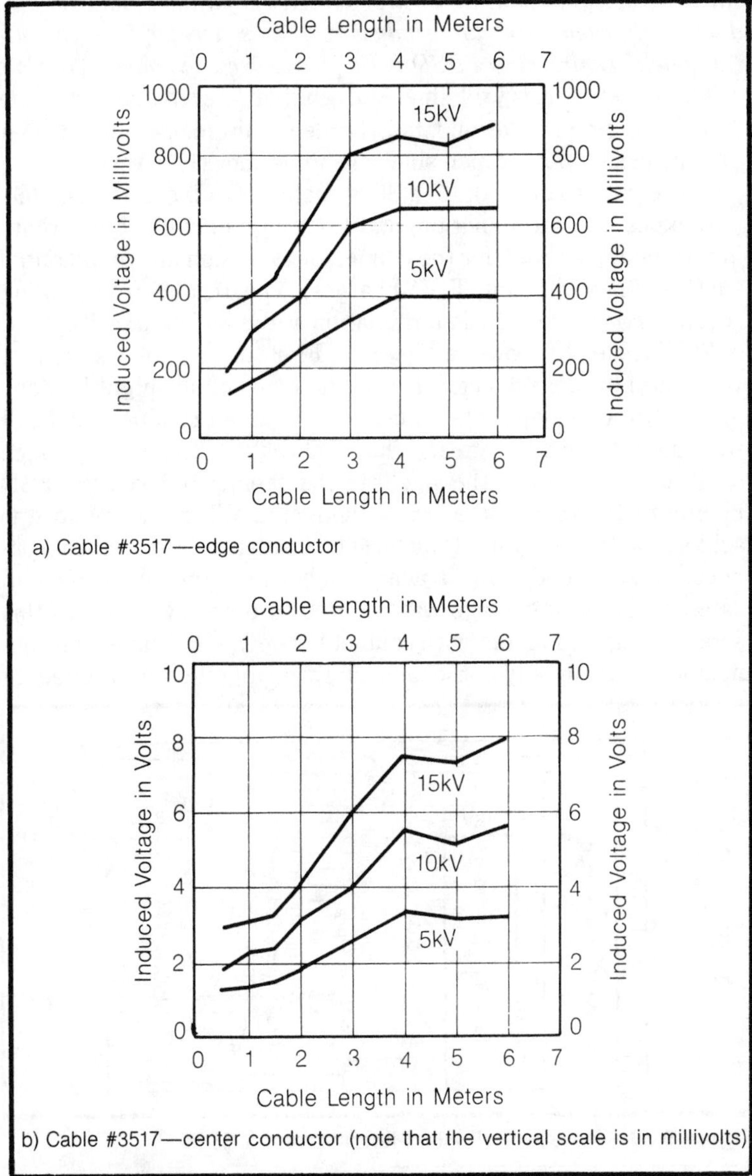

a) Cable #3517—edge conductor

b) Cable #3517—center conductor (note that the vertical scale is in millivolts)

Figure 5.35—ESD noise voltage as a function of cable length and discharge voltage.

shielded cables of identical construction, except for the number of conductors, the cable with more conductors (that is, a wider, more conductive shield) exhibited less ESD noise pickup than the narrower cable.

The benefits of a well shielded cable are never realized if the shield is improperly terminated. Three requirements must be met in a connector for proper shield termination: (1) a very low impedance path to ground; (2) a 360° metallic *hiding* of the conductors exposed for termination, and (3) a high enough conductivity in the connector body for the connector to remain an equipotential surface when carrying ESD charges. The results from testing several variations of shield termination at the *peripheral cabinet* end of 3517 cable are shown in Fig. 5.36. First, the shield was brought very close to the connector. It was then cut off, as might be done to avoid ground loops. The voltage on the center conductor of 3517 was estimated to be far greater than 500 volts.* Using a heavy gauge drain wire to connect the shield to the cabinet reduced the ESD pickup to 16 volts on the *center* conductor. When the shield was soldered to the miniature D metal shell insulation displacement connector and a solid contact achieved between the connector and cabinet, the noise voltage was reduced to only 1.25 volts. In this same configuration, but with contact between the cabinet and connector occurring only at the jackscrews as might occur after aging

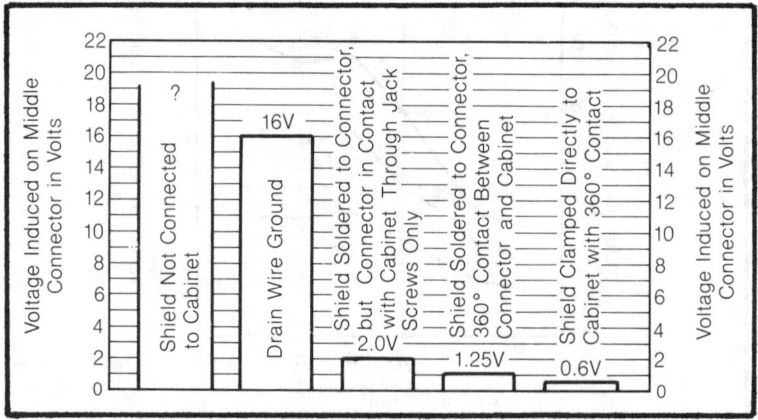

Figure 5.36—The effect of shield termination on ESD

*After this test, the oscilloscope was returned to the manufacturer for repairs.

or cosmetic painting, the noise voltage increased to 2.0 volts. The lowest noise pickup, 400 mV, was obtained with the shield clamped directly to the cabinet with 360° contact using angle clamps.

The increase in ESD susceptibility when the shield was grounded at only one end or connected only through a drain wire was traceable to the concentration of electromagnetic fields at the gap in the shield. This concentrated field coupled an increased noise voltage into the unprotected portion of the cable exposed through the gap. The *nontermination* of the shield provided neither a low impedance path to ground nor a hiding of the conductors. Terminating the shield through a heavy gauge drain wire (braid strap) can provide a low impedance path to ground but not adequate hiding of the exposed conductors.

From these tests, it can be concluded that to provide optimal ESD resistance, noise sensitive circuitry should use the central conductors of a shielded flat cable. Care should be taken to minimize the shield resistance by (1) keeping the length of the cable as short as possible; (2) using cables with highly conductive shields, and (3) providing low resistance connections to ground. When terminating the cable shield, providing a 360° metallic hiding of the exposed conductors at both ends of the cable is essential. Unshielded and ground-plane cables, provide little or no ESD protection.

One of the most remarkable aspects of Palmgren's findings is that, just as for coaxial cables (or generally speaking, any shielded envelope), until the shield is integrally bonded to the frame, what is seen is not the quality of the shield itself, but rather *to what extent the shield is ruined* by the more or less detrimental bonding impedance. We could say that unless the shield is integrally bonded, the shield transfer impedance Z_t is overruled by the shield termination impedance (pigtail or other) which, for the rise times of concern, can be several orders of magnitude larger than Z_t. In fact, relying on the transfer impedance of shielded flat cable could be misleading if one adheres strictly to the Shelkunoff definition of Z_t. Since, in flat cables, the shield is not part of the intentional signal path, we should speak of the differential transfer impedance Z''_t or Z_{td}; i.e., the ratio of the differential voltage appearing between two wires of a multi-conductor cable to the current flowing on the protective shield (discussion and practical data on Z_{td} can be found in Refs. 20 and 32).

In the case of ESD, all the traditional recipes for cable shields ("thou will ground the shield at one end only to avoid ground loops"), which are perfectly justified at audio or low frequencies, become irrelevant. To reduce ESD pick-up, a cable *shield* must be bonded to each housing that it penetrates. The argument that a floated shield will not allow the ESD current to flow and, therefore, will avoid coupling does not stand a more thorough analysis, nor does it agree with the conclusions of Palmgren's study: The floated end of the shield would reach several hundred volts of common-mode voltage and would re-inject some of it by capacitive coupling (more precisely, transfer admittance) on the wiring.

If a system is vulnerable to low frequency ambient noise (like ground potential differences between cabinets) and must also run trouble-free in possible ESD conditions, the dilemma is the following (typical of analog or instrumentation links):

1) If the protective shield is grounded on one end (generally the receiver end), the shield will not create a parasitic ground loop and will perform ideally as a Faraday shield against a low frequency EMI which may be there all the time. However, it will leave the system unprotected (or maybe even make things worse) during a casual ESD event.

2) If the protective shield is grounded at both ends it will protect against ESD a few times a day or week, but it may create a permanent *ground loop* problem.

Determining which of the two evils is less detrimental requires a quantitative approach. A more drastic solution is to immunize the system against low frequency common-mode, using CM baluns, signal transformers, differential drivers/receivers, PCB floating, etc., and having a good quality cable shield bonded at both ends to machine frames.

Assuming that bonding will be done almost perfectly, cable shields which are relied upon for ESD immunity must be selected carefully. Figure 5.37 gives guidelines on relevant shield parameters.

If the external cables exposed to ESD are coaxial cables, the noise appears to be due to the ESD induced current coupling via the transfer impedance of the coaxial braid. (A full explanation of the transfer impedance is beyond the scope of this book. The reader can find details of this mechanism in Ref. 20 and in several works

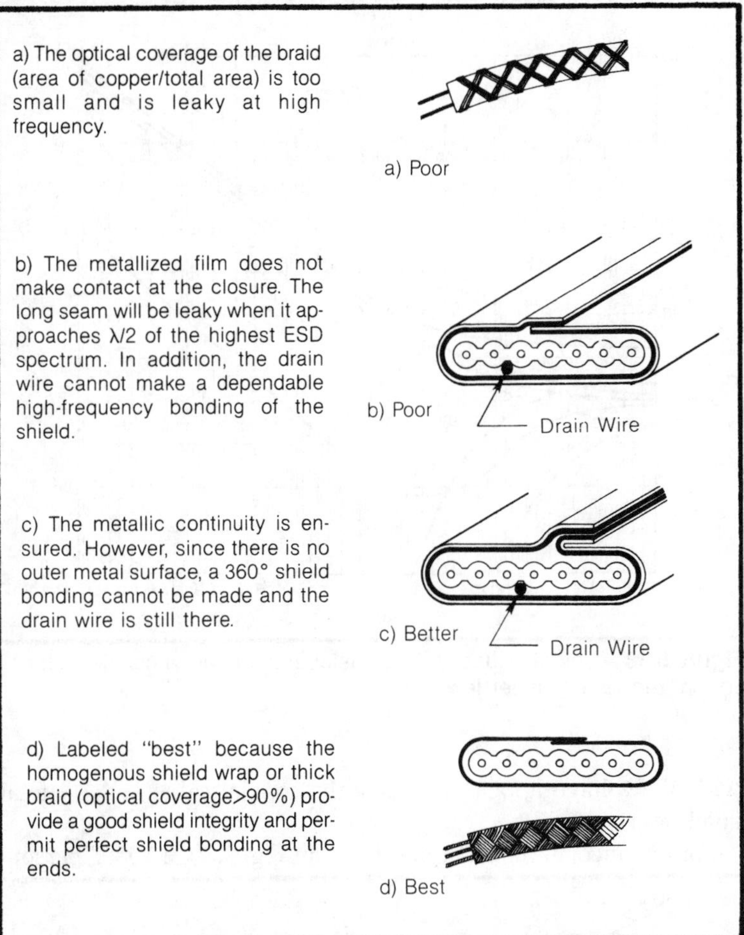

a) The optical coverage of the braid (area of copper/total area) is too small and is leaky at high frequency.
a) Poor
b) The metallized film does not make contact at the closure. The long seam will be leaky when it approaches λ/2 of the highest ESD spectrum. In addition, the drain wire cannot make a dependable high-frequency bonding of the shield.
b) Poor
Drain Wire
c) The metallic continuity is ensured. However, since there is no outer metal surface, a 360° shield bonding cannot be made and the drain wire is still there.
c) Better
Drain Wire
d) Labeled "best" because the homogenous shield wrap or thick braid (optical coverage>90%) provide a good shield integrity and permit perfect shield bonding at the ends.
d) Best

Figure 5.37—Cable shields aspects vs. ESD immunity

by E. Vance (Ref. 32). Figure 5.38 shows some typical values of the transfer impedance. This transfer impedance is defined as:

$$Z_{t\Omega/m} = \frac{V_i}{I_{sh}} \text{ per meter length}$$

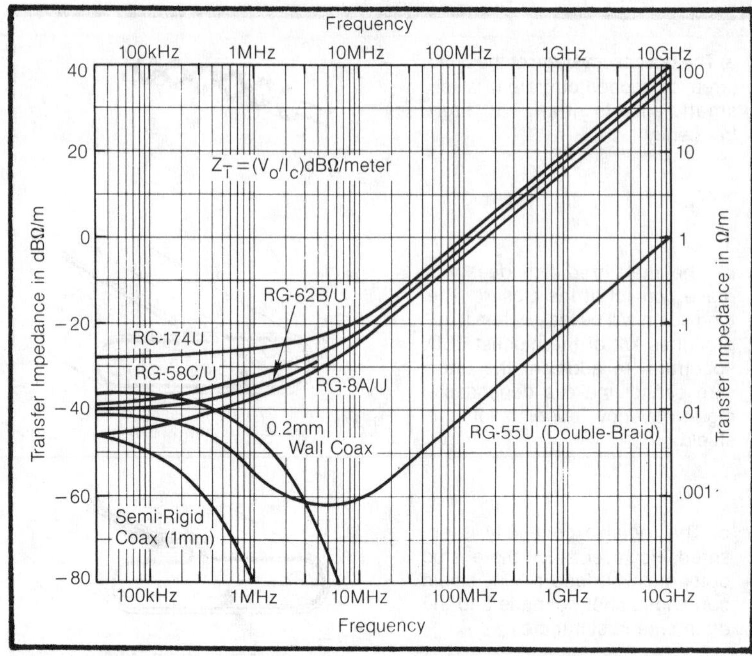

Figure 5.38—Typical values of the transfer impedance of coaxial cables, normalized to a 1 meter length

with V_i=voltage appearing *inside* the coaxial cable and I_{sh}=total shield current.

For instance, in the example shown in Fig. 5.39, a 10 kV person-

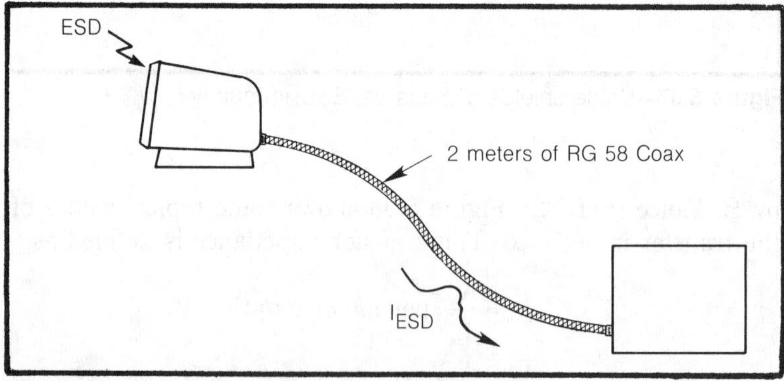

Figure 5.39—ESD example (external cabling)

nel ESD, with 1000 Ω source resistance, is applied to a video terminal. It has been found, using a current probe, that about half of the ESD current was returning to ground via the coaxial cable and the other machine.

The noise coupled inside the coax can be found by the transfer impedance:

- Portion of I_{ESD} flowing on coaxial shield:
  - —5 Amps peak, rise time τ_r=10 nsec (was slowed down by coaxial external inductance)
  - —2nd corner frequency=$1/\pi\tau_r \cong 30$ MHz
  - —Z_t@30 MHz, for RG 58=−10 dBΩ/m≡.3 Ω/m
- ΔV drop inside shield≂.3 Ω/m×2 meters×5 Amp=3 volts
- Solutions:
  - —Use cable with lower Z_t (triax, reinforced braid, etc.)
  - —Use double shield with outer shield grounded to box
  - —Add ferrite *over* the coax to increase its external inductance

Often overlooked in their contribution to ESD coupling are the power cords; since they connect to the transformer or other bulky components, they seem beyond suspicion. However, as it penetrates the machine enclosure, a power cord carries internally the transients induced by ESD. In this case, the solution lies in filtering/decoupling the power cord right at its point of entry. Since power cords are generally filtered anyway, this seems a superfluous recommendation, but RFI filters are generally optimized to meet conducted specifications below 30 MHz and may be inefficient at the 100 MHz (or greater) range of the ESD spectrum. Thus, the filter could be improved by additional ferrites, or it could be mounted in such a way that its input-to-output isolation is still efficient for ESD rise times.

Another solution, not mutually exclusive, is to use a shielded power cord. A shielded power cord will both avoid ESD re-entry and also improve the drain path to ground.

Finally, when an installation with unshielded cables has ESD problems, instead of replacing all cables with shielded ones, which can be the long-term answer, a field fix can be installed by using a zip-on shielded jacket over the whole bundle (Fig. 5.40).

For any of the shielded cables which have been discussed, the bonding of the cable shield to the case is crucial since it provides the transition from cable shield to housing integrity. Direct bonding of cable shield to the *outer skin* of the box will prevent ESD

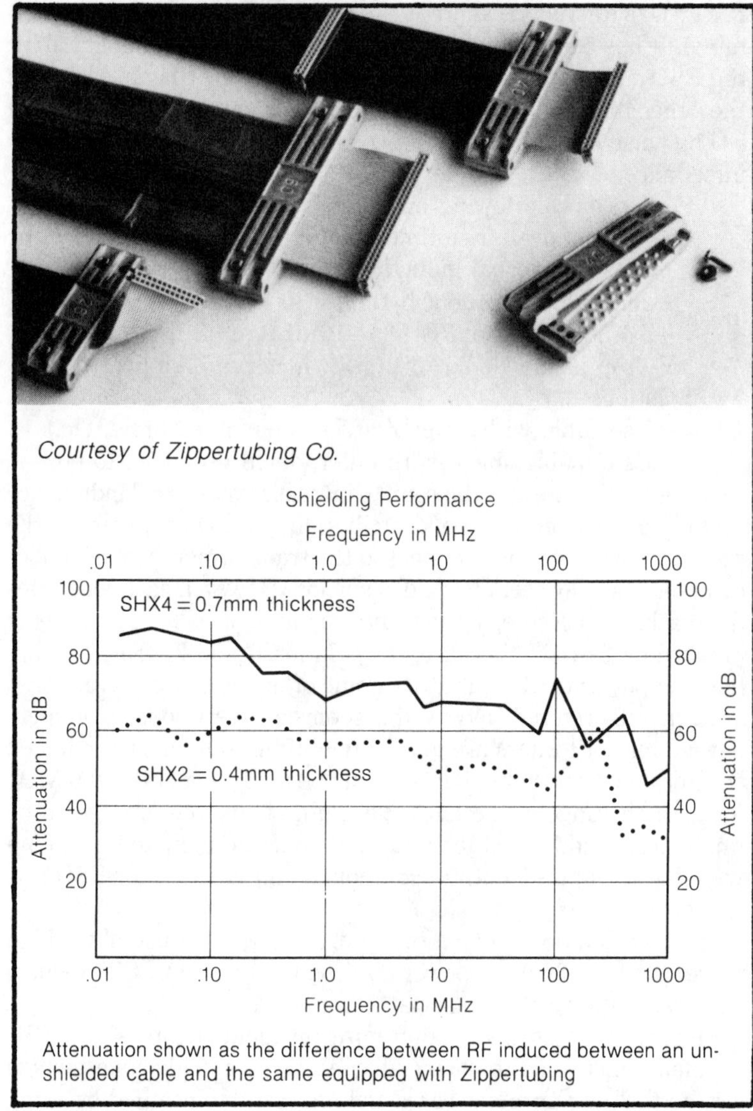

Figure 5.40—Shielded zip-on jacket. The shield is made of knitted wire mesh (copper clad steel, tinned) with a grounding braid. A sturdy mounting clamp, providing a full circumferential bonding, is available.

current from the shield from re-radiating inside (Fig. 5.41). To emphasize this need, Fig. 5.42 shows the drastic improvement when a metallic continuity is ensured from cable shield-to-connector-to-box. Several ways of achieving this continuity are shown in Figs. 5.43, 5.44 and 5.45. A problem arises when, depending on the system configuration and the type of peripheral units which are connected, some connectors are temporarily left without cables. Since it is very likely that the inner wiring to the electronics has been installed, these unemployed connectors are an invitation to serious ESD problems if they are accessible. If a bare pin can be approached by a finger at less than 1 cm or so, a direct discharge can be conveyed straight into the sensitive components. To avoid this the designer should either:

- not provide the inner wiring when the corresponding peripheral is not installed
- mount the connector in a recessed area so that a would-be charged finger will zap the housing first, or
- provide a metallic cap or blinder when the connector is not used.

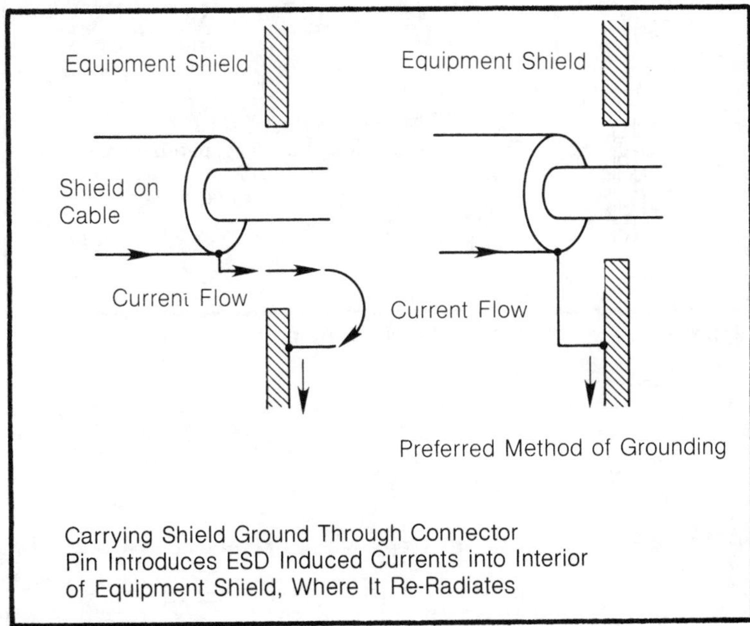

Figure 5.41—Grounding of cable shield for ESD excitation

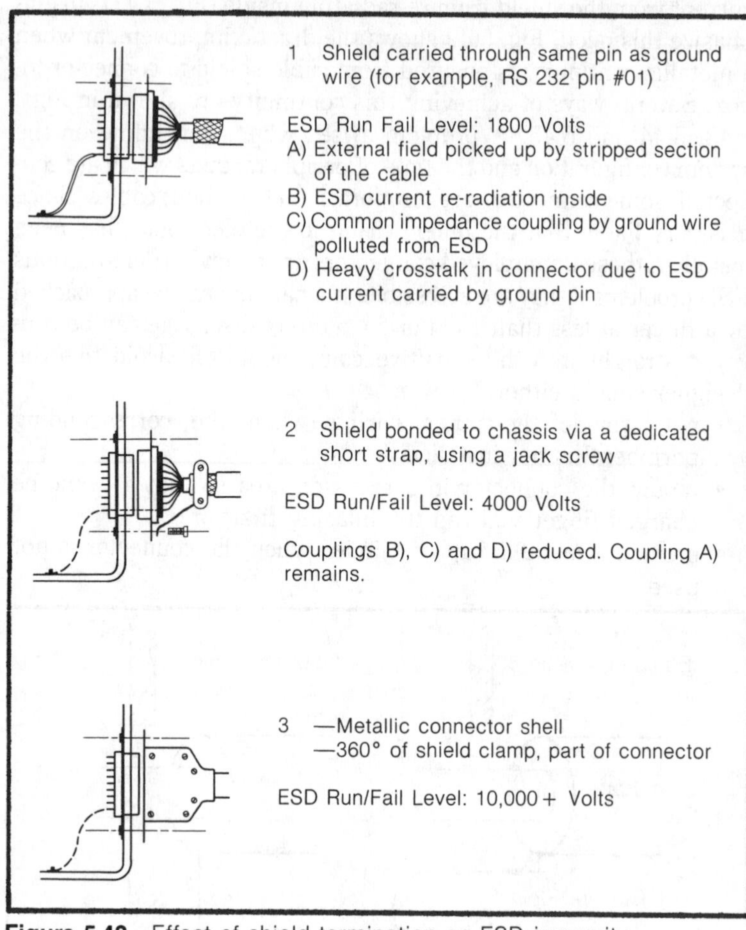

1 Shield carried through same pin as ground wire (for example, RS 232 pin #01)
ESD Run Fail Level: 1800 Volts
A) External field picked up by stripped section of the cable
B) ESD current re-radiation inside
C) Common impedance coupling by ground wire polluted from ESD
D) Heavy crosstalk in connector due to ESD current carried by ground pin
2 Shield bonded to chassis via a dedicated short strap, using a jack screw
ESD Run/Fail Level: 4000 Volts
Couplings B), C) and D) reduced. Coupling A) remains.
3 —Metallic connector shell
—360° of shield clamp, part of connector
ESD Run/Fail Level: 10,000 + Volts

Figure 5.42—Effect of shield termination on ESD immunity

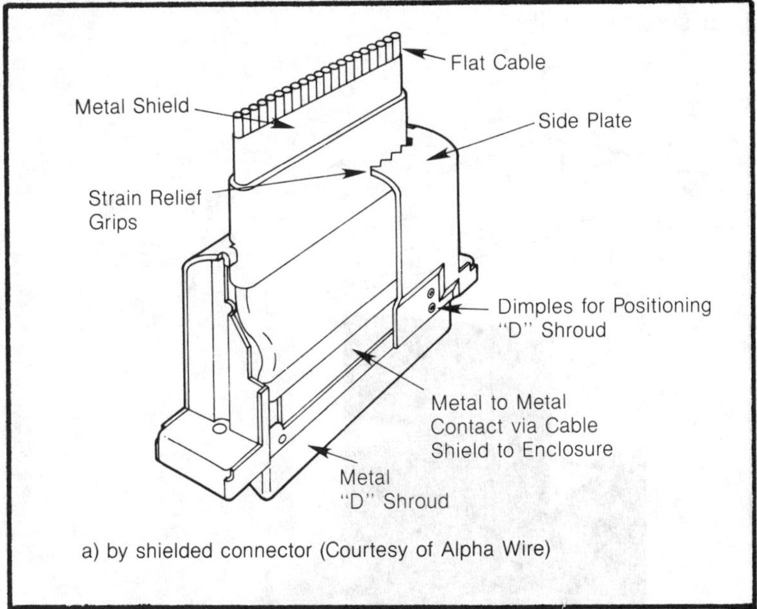

a) by shielded connector (Courtesy of Alpha Wire)

Figure 5.43a—Proper bonding of flat cable shields

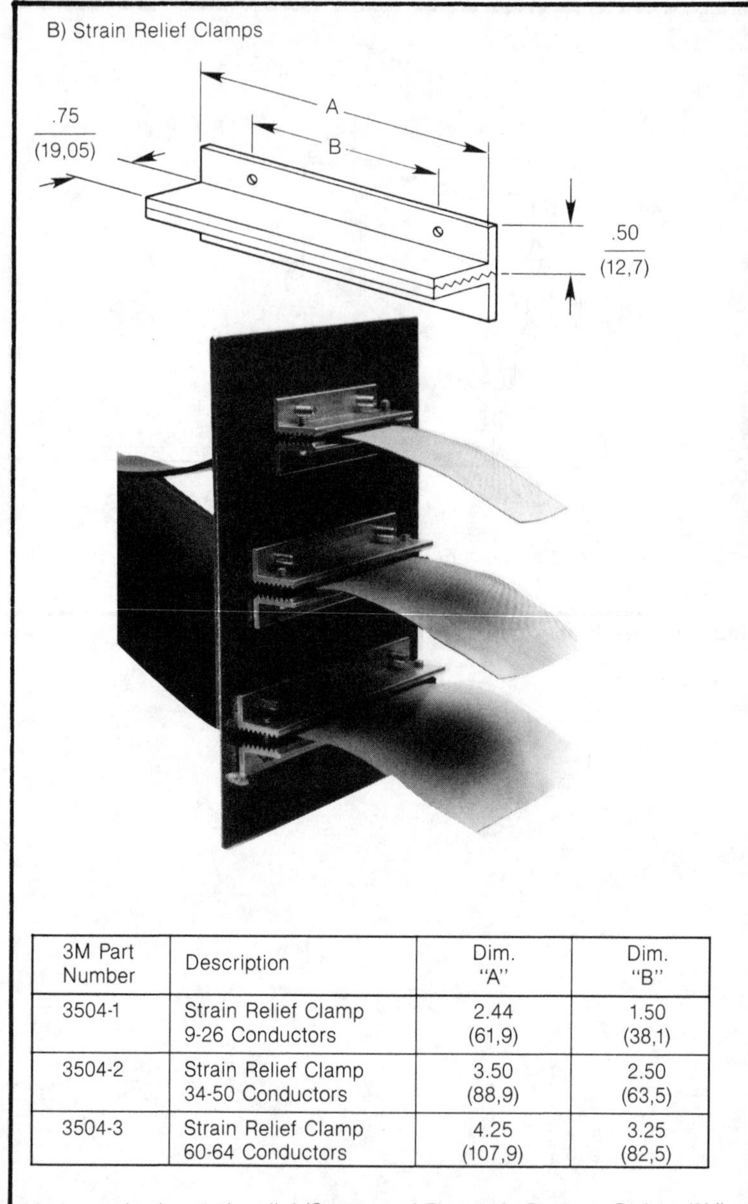

3M Part Number	Description	Dim. "A"	Dim. "B"
3504-1	Strain Relief Clamp 9-26 Conductors	2.44 (61,9)	1.50 (38,1)
3504-2	Strain Relief Clamp 34-50 Conductors	3.50 (88,9)	2.50 (63,5)
3504-3	Strain Relief Clamp 60-64 Conductors	4.25 (107,9)	3.25 (82,5)

b) via conductive strain relief (Courtesy of Electronic Products Division/3M)

Figure 5.43b—Proper bonding of flat cable shields

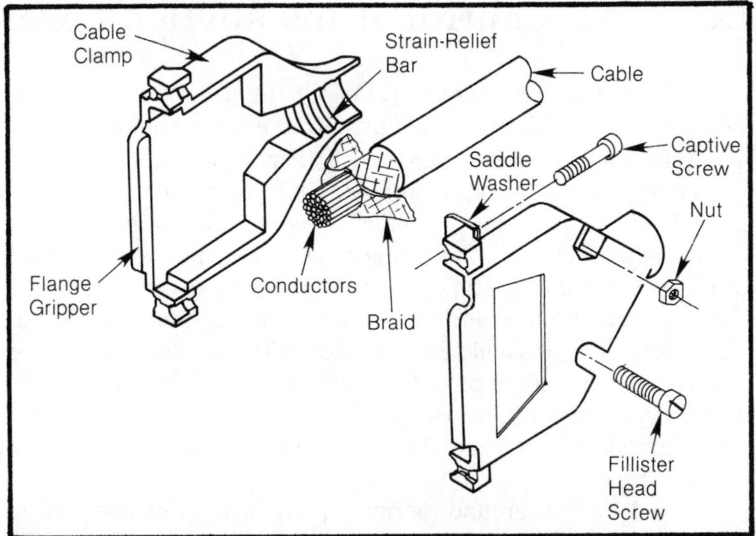

Figure 5.44—Conductive connector backshell with integral braid retention (from AMP, Inc.)

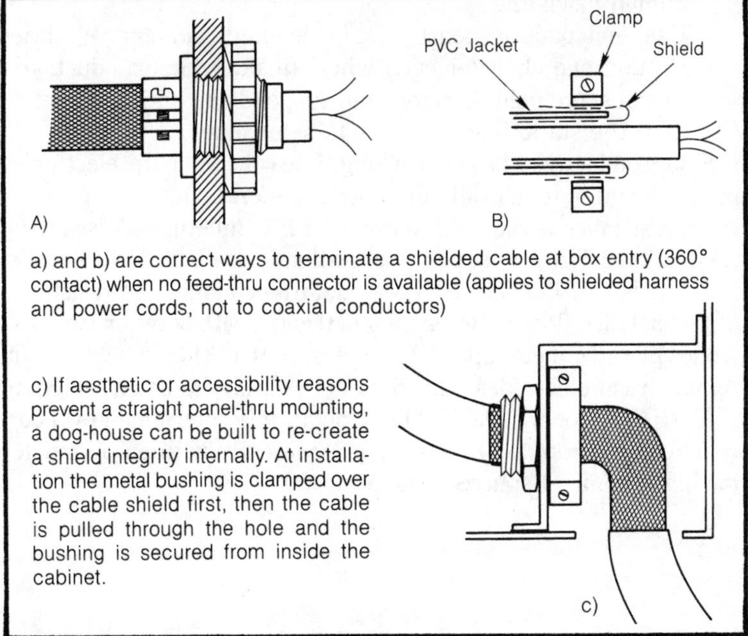

Clamp
PVC Jacket
Shield
A)
B)
a) and b) are correct ways to terminate a shielded cable at box entry (360° contact) when no feed-thru connector is available (applies to shielded harness and power cords, not to coaxial conductors)
c) If aesthetic or accessibility reasons prevent a straight panel-thru mounting, a dog-house can be built to re-create a shield integrity internally. At installation the metal bushing is clamped over the cable shield first, then the cable is pulled through the hole and the bushing is secured from inside the cabinet.
c)

Figure 5.45—Shielded cables terminations without connectors

5.7 ESD Control at Installation Level

Installation environment is probably the aspect of which the designer has the least control. Furthermore, since equipments are generally specified for certain environmental conditions, it would be foul-play to change the environment because an electronic equipment does not meet the challenge! However, life seldom provides clear situations like this. Many times the environment was unknown or poorly defined at the time of initial equipment design. Moreover, salesmen might have overdone it and sold an equipment for an environment where it should not have been installed. So, if everything else fails, or at least to provide temporary relief while waiting for a more engineered solution,

- Maintain relative humidity above 50%, or use an ionized air blower
- Avoid carpet around perimeter, or use grounded woven metal thread in carpet, or cover carpet with metallized mats, or use anti-static sprays on carpet, or use grounded rail for human discharge
- Use seat pads in nearby chairs having *breather-type* fabrics
- Ground one chair (or cart) wheel (make wheel conductive)
- Install static field detectors and suspend any critical operation until the static field resumes to safe values.

Floor conductivity has been a long-time concern in the electronics industry and in industrial buildings in general. For instance, electrical resistance is required to meet NFPA Bulletin 56A issued by the National Fire Protection Association in the U.S. or, in Europe, DIN Std No. 51953. The NFPA standard recommends a resistance of the installed floor between 25,000 Ω and 1 MΩ, between two electrodes placed 1 meter apart. The measurement should be made via 2 cylindrical electrodes with 6.35 cm (2.5 in) diameter, weighing 2.25 kg (five pounds) each. The resistance is determined by a current reading from a 500 V source, or directly with an ohmmeter having a nominal source voltage of 500 V.

Chapter 6

ESD Case Histories

This chapter relates some ESD "war stories" experienced mostly by the author himself or by his associates at an interference control consulting firm in the U.S. There are periods when the consultants' telephone rings several times a day just for ESD calls. Assisting a customer who has an ESD problem is seldom a boring, "déjà vu" experience. Of all the EMI manifestations, ESD is probably the one whose symptoms can be the most varied and deceptive and whose diagnosis can be the most elusive. Needless to say, ESD is also a privileged area for Murphy to exercise his laws with demoniac ability.

The following tales from the trenches have been selected because they share several similarities:

- In general, the plaintiffs were knowledgeable engineers who had tried all common-sense fixes (plus a few others) before calling a consultant.
- The outcome of these fixes (whether or not they helped) usually was not documented, or at least not quantitatively. When they were, it was in the form of a verbal legacy passed on by each of the frustrated raiders to his next partner. The GO/NO-GO levels reached with each fix, the manufacturer type and part number of the suppression components, and whether the fixes were cumulative or if each successive fix was taken away before a new one was tried could not be determined. The type of simulator used was sometimes not even documented. The whole saga was summarized as: "We have tried everything, and nothing works."
- Sometimes, the client had already called a consultant who tried several pieces of the usual EMC arsenal, including in one case a complete overhaul of the building earthing network, ground

rods, etc. which, although it certainly improved the safety of the facility, did not do much for the ESD immunity of the machine.

- In several instances, many of the unsuccessful fixes *were, in fact, appropriate.* But they were tried in random sequences so that the engineers could not compile and interpret the gradual changes they might otherwise have noticed, thereby not learning from their mistakes.
- Finally, there was unanimous consent about what the consultant had to do: He was sentenced to fix in 48 hours what three successive task forces (appointed by four consecutive Division Managers) had failed to solve in eight months.

6.1 Case #1—The Re-radiating Ground Strap

The case of the re-radiating ground strap started as an easy one. The EUT was a process control mainframe housed in a steel cabinet. The CPU was failing at 5 kV with an IEC-type gun, while ESD immunity of at least 12 kV was desired. A short discussion over the phone revealed that the operator panel front cover had not been grounded. However, grounding it by a few short straps to the main housing did not improve performance by more than 1 kV. A quick trip to the customer's site revealed that the problem had, in fact, three facets as shown in Fig. 6.1.

- The flat cable going from the operator display to the microprocessor was run tightly against one of the cover ground straps. During an ESD event, this ground strap was a preferred sink path for the discharge current to ground. This created a strong magnetic field around the strap which coupled into the ribbon cable conductors.
- To make things worse, the zero-volt return of the operator display board was made via a heavy wire, part of the dc-supply harness, but it was running quite far from the flat cable. Although a zero-volt wire was also provided within the flat cable, a large ground loop existed between the signal cable and the dc-0volt return.

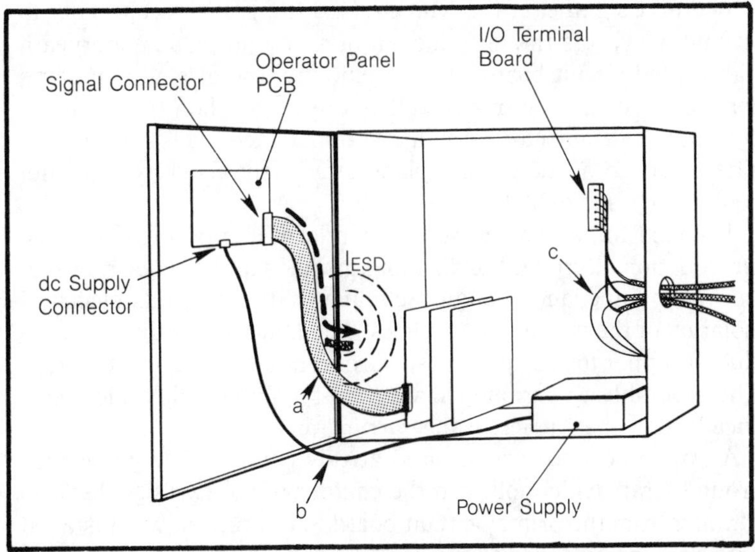

Figure 6.1—Case History #1. Packaging of the machine shows a) the signal cable routing, b) the dc supply wiring including the OVolt, and c) the I/O cable pigtails

- On the opposite side, the I/O cables were entering the cabinet through a simple hole. The shields of each pair were terminated by long (in terms of ESD wavelength) pigtails collected by a grounding post *inside* the frame.

The design of all PCB cards was quite clean, so as soon as the first two problems were corrected, the level went up to 12 kV. A change in the I/O cables was recommended to bring the level even higher.

6.2 Case #2—ESD Hardening of a Typewriter

A typewriter prototype had to be hardened to 8 kV minimum. Since the users could have either metal desktops or non-conductive desks, the criteria had to met for both configurations. Initially, the EUT failed at approximately 4 kV using the typical IEC-801 test method.

The first significant fix, which raised the ESD susceptibility to around 8 kV, was the introduction of a ground plane underneath the printed circuit board. At this juncture, the ground plane was connected to the power zero-volt reference on the printed circuit board at the connector using a low impedance, flat-strap jumper. The presence of the ground plane effectively acted as a low impedance Faraday shield.

The next fix, which resulted in significant improvement, was the introduction of proper bonding and grounds to the typewriter carrier. Low impedance bonding eliminated the previous electrical isolation of the carrier components. The bonding configuration included copper foil straps between the front and back of the carrier. The assembly was grounded via a metal sliding clip which electrically tied the carrier to the carrier rail.

A low impedance was desired at this point. A low impedance ground strap, which followed the contour of the two signal cables running from the printed circuit board to the carrier, was installed with one end soldered to the carrier bonding strap and the other end soldered to the ground plane. This last fix also provided a ground plane and shielding for the two signal cables against the effects of field coupling, as well as providing for low impedance carrier grounding. Also included in the grounding scheme was the establishment of electrical continuity between the platen frame and the carrier rail frame via straps mounted at the ends of the carrier rail (see Fig. 6.2).

At this point the typewriter's susceptibility had been raised to 12 kV with the typewriter placed directly upon the large table ground plane. When the unit was placed upon a wooden box, thus elevating it approximately 10 inches above the ground plane, the susceptibility dropped to about 7 kV. An additional two fixes increased the susceptibility of this second test configuration to 12 kV. The most significant of these was the placement of a shield directly underneath the keyboard unit. The best results were obtained by tying this shield to the carrier rail. The other fix involved the power supply cables, which ran from the rear of the typewriter to the printed circuit board. These cables, which carried the dc supplies to power the logic, motors and solenoids, were twisted tightly together as a single bundle. At this point, the overall typewriter ESD susceptibility was raised to 12 kV.

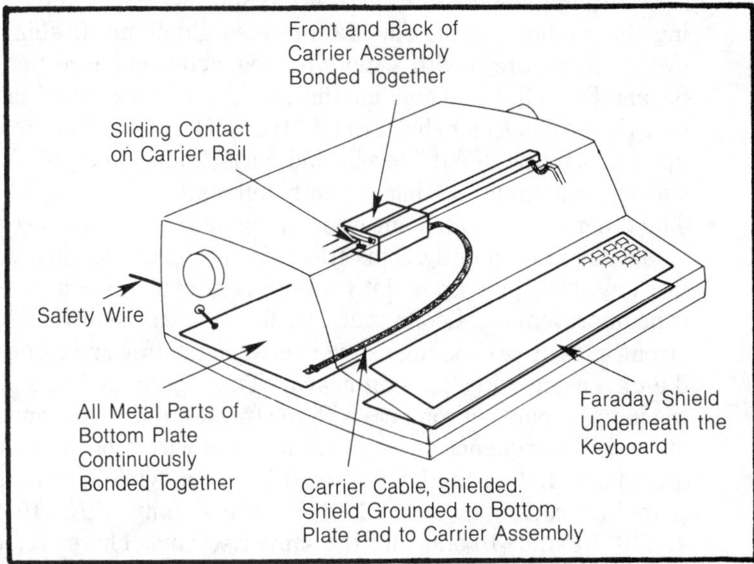

Figure 6.2—Hardened typewriter of Case History #2 (>8 kV ESD)

6.3 Case #3—The Data Terminal with Floating Tray

This data terminal was a more challenging mission. A significant amount of hardening had already been attempted by the designers with some success. The unit was withstanding 7 kV and failing at about 8 kV, but a level of 20 kV was desired. As in most ESD hardening problems, the last steps of the ladder are the most difficult to climb.

After several investigations, including carefully tracking the results, *good or bad,* of the previous attempts, attention was centered on two major failure modes:

- Although a conductive coating had been applied inside the terminal, a paint overcoat was sprayed in the exterior of the box. Nobody had paid special attention to the edges, and they were covered with the decorative paint so that no electrical continuity existed along the seams (Fig. 6.3). Even worse, this was the very point where the data bus cable entered the box.

Restoring the metallization over the abutting edges, replacing the mediocre cable shield (one-sided aluminum flashing over a mylar wrap) with a more homogeneous one (see Sec. 5.6 and Fig. 5.35), and bonding this shield to the box metallization via the connector shell brought the "GO" level in this area up to 20 kV (the "FAIL" level could not be found since 20 kV was the maximum setting of the simulator.)

- The other susceptible point was a metallic tray that the user would touch frequently. This tray was originally floating. A previous attempt to ground it via a jumper did not give significant improvement. Looking at Fig. 6.4, one can see why: A strong capacitive coupling existed between the tray and a bundle of sensing wires passing nearby. The high impedance of the jumper could not compete with the tray-to-cable capacitance at the ESD frequencies. Ten centimeters of wire represent an impedance L/dt=100 nH/10^{-9} sec=100 ohms, while a typical parasitic capacitance of 10 pF offers only dt/C=10^{-9} sec/10^{-10} Farad=10 ohms for the same rise time. Therefore, a significant percentage of the ESD current was coupled into the sensor harness. By grounding the tray via a short, wide strap, and by putting ferrite beads over the sensing wires, the situation was reversed; i.e., most of the current was now sinking to ground by the metallic coating. This improvement, too, brought the level in the tray area up to 20 kV.

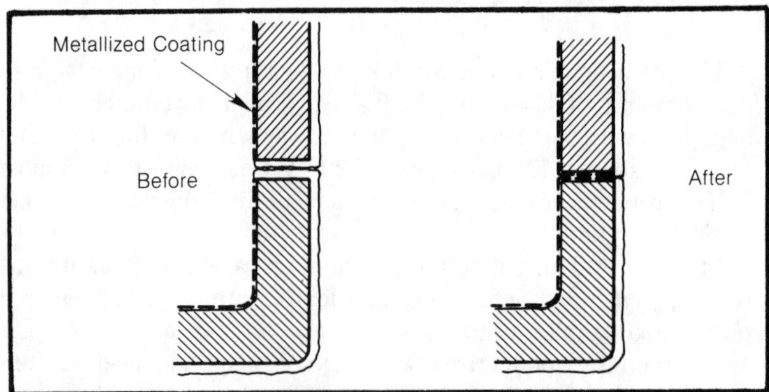

Figure 6.3—Case History #3. Exaggerated view of the problem caused by paint sprayed over the metallized coating, preventing any electrical contact

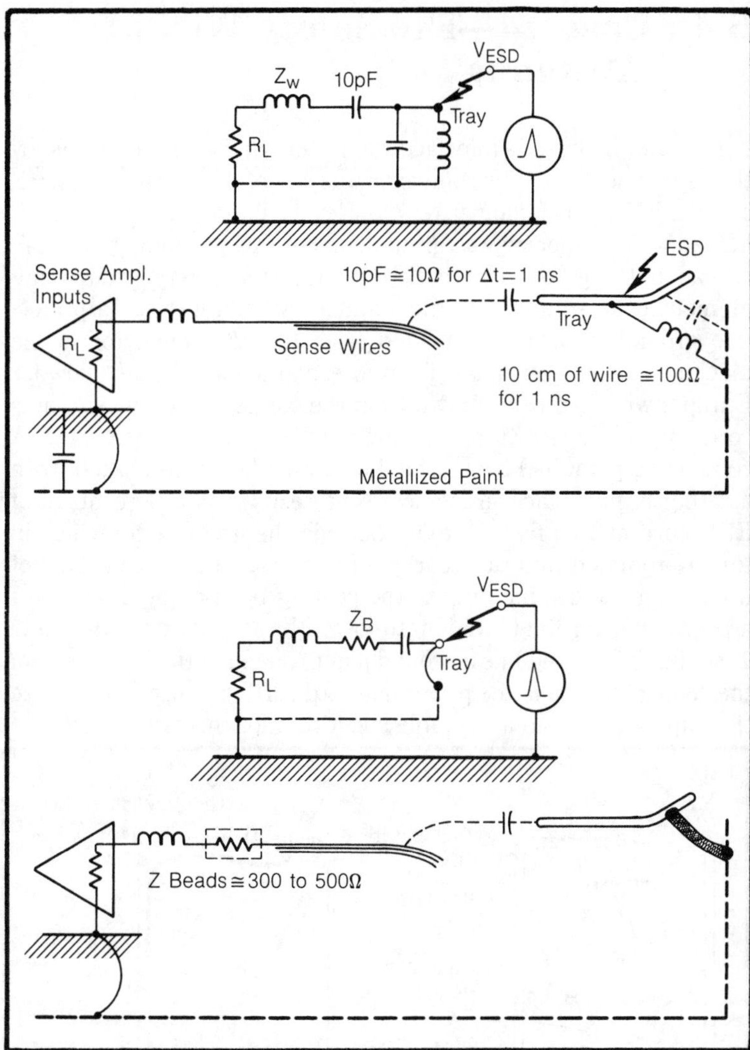

Figure 6.4—Case History #3. Figure a) shows the coupling path before the tray was closely bonded to metallized housing. Figure b) shows the coupling circuit after the tray was tightly grounded and ferrite beads were added.

6.4 Case #4—The Safety Wire "Antenna"

This unit had a data interface connector which was accessible by the user. Therefore, this connector was one of the discharge points, and the EUT was failing at very low levels. Figure 6.5a shows that, indeed, this connector was grounded. But this grounding originally had nothing to do with ESD and was a safety ground implemented to comply with the standard stipulating that a protective ground was to be provided from the I/O connector to the machine earthing terminal. Therefore, a green (or green/yellow for Europe) wire had been drawn from the connector to the ac mains cord earthing post. Moreover, the machine had a plastic case. A conductive paint had been used, but it was the graphite kind, with a rather high surface resistance. As a result, some percentage of ESD current was flowing to ground via the green wire, which in turn re-radiated into the nearby electronics. The fix consisted of upgrading the conductivity of the coating by sticking a wide 3M adhesive copper foil (with length-to-width ratio not exceeding 5 to 1, so that it behaved like a ground plane) on top of the coating from the connector area to the power line cord earthing post. Then, since this tape was not an acceptable substitute for the safety wire, the

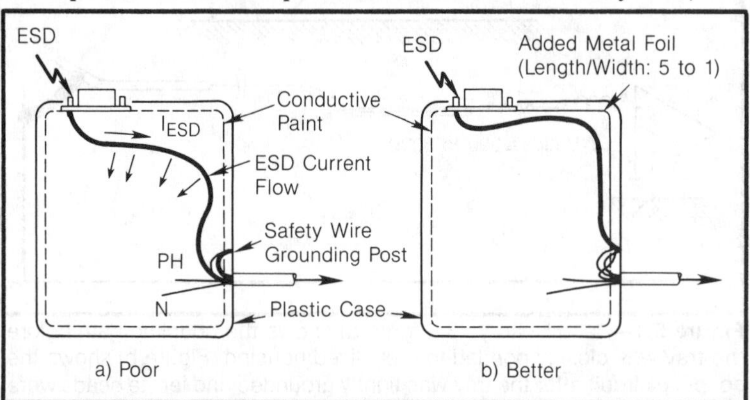

Figure 6.5—Case History #4, before and after the fix

wire itself had been kept, but re-routed tightly against the copper foil. As a result:

- most of the ESD current (high frequency) was now flowing into the foil
- the rest of the ESD current which still came by the safety wire was not causing re-radiation because the "antenna" was laid on a ground plane.

6.5 Case #5—The Trigger-happy Watchdog

The unit in this case was a microprocessor for industrial process control. The whole unit consisted of two cards mounted side by side in a case. Although originally a metal can had been designed, the client wanted to meet the challenge of 8 kV ESD with a plastic housing instead. In this case, of course, the test was performed with indirect ESD. Initially, the unit was failing at 3-4 kV on practically all sides. An examination of the PCB (using drawings) revealed about 12 undesirable trace loops, all of which were patiently eliminated using 3M copper tape to fill the voids or heavy-up the grounds. After this, the whole card sustained 8 kV successfully, except in one very critical area which was reluctant to show any improvement. The irritating aspect was that absolutely no critical circuit existed in this area. In discussing the problem with the designer, the culprit was finally caught: To save volatile data during transient power loss, a power sensing circuit had been designed. This sensor was recognizing a "power-loss condition" every time the bulk, unregulated dc voltage dropped by more than 10%. The time constant had to be fast enough so that, once a power loss was declared, there was still 10 msec left (about 1/2 cycle, or the time during which a voltage regulator still can make up for the failing power mains) before the 5 Vdc actually dropped below 4.5 V. During that time, using a "watchdog" procedure, a STORE signal was sent and the volatile data were saved into a non-volatile memory. When power was restored, a RECALL signal was sent to the nonvolatile RAM and the microprocessor resumed its previously interrupted operation. Unfortunately, the circuit was on the edge of the card and had no ground plane underneath it. During ESD conditions, induced glitches would foul up the power-loss and store/write commands, creating false SAVE routines. The solution was to run the sense traces close to a ground land on the PCB and to decouple the circuit slightly so that it would not respond to glitches of a few nanoseconds but would still match the 100 ns window of the TTL store signal.

6.6 Conclusion: Troubleshooting Hints

Armed with the explanantion of ESD coupling mechanisms given throughout this book, the reader can identify most of the ESD failure modes and apply the proper fixes. To help quantifying the improvements, and to avoid false routes, the following guidelines have proven to be useful:

- Always grade the progress brought by a modification in terms of the new GO/NO-GO level or in terms of the new error-per-pulse ratio (see Sec. 4.5).
- Try to visualize the ESD current paths and how they are modified by the bonding/grounding changes. This can often give a clue to the validity of a fix.
- To monitor either the ground currents or the common-mode current induced in cables, an RF current probe is a priceless tool. Being a totally floated sensor, it is not affected itself by ESD, and it does not modify the victim circuit.
- As an *absolute rule, never remove a fix* when, although it seemed to make sense, it did not bring any significant improvement. ESD can couple into the victim by many parallel paths and until

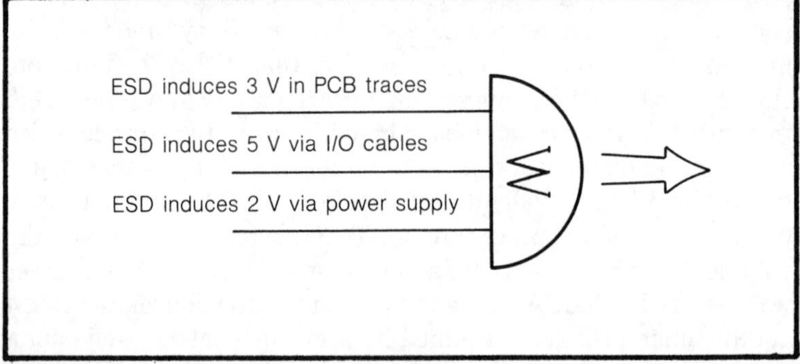

Figure 6.6—Conceptual view of the simultaneous effect of more than one ESD path

ALL of them have been reduced, a change may not be noticeable. For instance, imagine an equipment where, as shown in Fig. 6.6, the transients induced from a given ESD pulse arrive on the victim circuit simultaneously by 3 different paths:

—3 volts are induced in PCB traces
—5 volts are picked up by external signal cables
—2 volts come in via the power supply

If the victim sensitivity is ≅1 volt for a pulse of a few nanoseconds, each one of these paths is enough by itself to upset the circuit. Therefore, if someone has a bright idea and tries to filter the I/O cable, he may reduce this particular contribution to 0.3 V and not notice significant progress because the other two paths are still there. He could then conclude that his fix was useless and take it away to try something else. Dozens of fixes can be tried this way and will never work!

If the engineer had left the first fix in place, then tried a fix on the PCB which reduced the induced noise to 0.2 V, he would have noticed a slight improvement, but not the expected amount. Finally, when the third coupling is also reduced say, down to 0.1 volt, a sudden improvement shows up because a one order of magnitude hardening has just been achieved.

Appendix A

ESD Protection by Design of Chips and Microcircuits

The drastic reduction in area and thickness of the active elements occurring in recent generations of semiconductors makes each new logic family more prone to ESD damage than the former one. Technologies achieving more than 1,000 gates/mm² and offering speed-power products much less than a picojoule, with propagation delays inferior to 100 picoseconds, will have channel length of .25 microns. This corresponds to a theoretical breakdown voltage of only 20 volts!

Designers of integrated circuits, hybrids and microelectronics can generally build in a certain level of ESD hardening through layout precautions and integrated protection networks. This hardening will make the chips or modules safer to handle and will relieve the user of some of the cost and burden of hardening. The following guidelines are taken from Military DOD-Handbook 263 and from technical notes from manufacturers such as National Semiconductor and Texas Instruments.

A.1 Protection of CMOS Devices*

General. Various protection networks have been developed to protect sensitive MOS. These circuit protection networks provide limited protection against ESD. Many of the protection networks designed into MOS devices reduce the susceptibility to ESD to a maximum of 800 volts. MIL-M-38510 V—ZAP test voltages for CMOS, for example, vary from 150 to 800 volts. Protection circuitry of some

*Section A.1 is taken from MIL-HDBK-263

devices is improving and protection to 4,000 volts appears to be achievable for some MOS devices. However, electrostatic potentials of tens of thousands of volts can be generated in uncontrolled environments.

A.1.1

The protection afforded by specific protection circuitry is limited to a maximum voltage and a minimum pulse width. ESDs beyond these limits can subject the part's constituents to damage or damage the protection circuitry constituents themselves which are also often made of moderately or marginally sensitive ESDS parts. Damage to the protection circuitry constituents could result in degradation in part performance or make the ESDS part more susceptible to subsequent ESDs. The degradation, for example, could be a change in speed characteristics of the ESDS parts or an increase in leakage current of the ESDS part. Multiple ESDs at voltages below the single ESD pulse sensitivity voltage or energy level can also weaken or cause failure of the part performance or protection circuitry constituents resulting in degradation or failure of the ESDS part. Loss of protective circuitry may not be apparent after an ESD.

In summary, protection networks reduce but do not eliminate the susceptibility of a part to ESD. This reduction in ESD sensitivity, however, results in a lower incidence of ESD part failure.

The sensitivity of the same type of ESDS part can vary from manufacturer to manufacturer. Similarly, the design and the effectiveness of protection circuitry also varies from manufacturer to manufacturer.

A.1.2 Design Precautions

Various design techniques have been employed in reducing the susceptibility of parts and assemblies to ESD. Diffused resistors and limiting resistors provide some protection, but are limited in the amount of voltage they can handle. Zeners require greater than 5 nanoseconds to switch and may not be fast enough to protect an MOS gate. Furthermore, zener schemes, diffused resistors and limiting resistors reduce the performance characteristics of the part which in many instances are the primary consideration for which that part was designed.

A.1.2.1 Parts and Hybrids Design Considerations

Some design rules to reduce ESD sensitivity for parts and hybrids are as follows:

a. MOS protection circuitry improvement techniques are: increasing diode size; using diodes of both polarities; adding series resistors;

and utilizing a distributed network effect;

b. Avoid cross-unders beneath metal leads connected to external pins; otherwise treat the part as electrostatic sensitive. Also, since cross-unders are diffused during the N+ (emitter) diffusion process, the oxide over the diffusion will be thinner, causing this area to have a lower dielectric breakdown. Deep N diffusions, rather than N+ diffusions, should be used for cross-unders, if a deep N diffusion step is used in the fabrication process;

c. MOS protection circuits should be examined to see if the layout permits the protection diodes to be defective or blown without causing the circuit to be inoperative;

d. Distance between any contact edge and the junction should be 70 microns or greater on bipolar parts;

e. Linear IC capacitors should be paralleled by a PN junction with sufficiently low breakdown voltage;

f. For bipolar parts avoid designs permitting a high transient energy density to exist in a PN junction depletion region under ESD. Use series resistance to limit ESD current or use parallel elements to divert current from critical elements. The addition of clamp diodes between a vulnerable lead and one or more power supply lead can improve ESD resistance by keeping critical junctions out of reverse breakdown. If a junction cannot be kept out of reverse breakdown, physically enlarging the junction will make it more ESD resistant by reducing the initial transient energy density in inverse proportion to its area;

g. The protection of a transistor from ESD can be improved by increasing the emitter perimeter adjacent to the base contact. Enlarging the emitter diffusion area also helps in some pulse configurations;

h. As an alternative to using clamping diodes, which consume chip area and can cause unwanted parasitic effects, a "phantom emitter" transistor can be used to improve ESD resistance. The phantom transistor incorporates a second transmitter diffusion shorted to the base contact. This creates a deliberate separation of the base contact from the normal emitter without interfering with normal transistor operation. The second emitter provides a lower breakdown path BV_{CEO} between the buried collector and the base contact;

i. Avoid pin layouts which put the critical ESD paths in corner pins which are prone to ESD;

j. Avoid metallization cross-overs where possible. These cross-over areas are typically separated by thin dielectric layers. Cross-overs often impose a number of metallurgical requirements which are frequently incompatible. For example, once the first metallization layer (Al) is deposited, the circuit cannot be subsequently heated in excess of 550°C because the eutectic point of Al-Si system is 575°C. Thus, the dielectric layer (SiO_2) should be deposited by a low temperature process such as pyrolytic deposition. This layer is prone to breakdown from ESD for two reasons:

(1) A low temperature growth of SiO_2 generally is not uniform in thickness and not free from pin holes;

(2) The dielectric layer is thin and thus the breakdown voltage is very low;

k. Avoid parasitic MOS capacitors whenever possible. Microcircuits with metallization crossing over low resistance active regions, that is, V_{CC} over N+ isolated diffusions, are moderately sensitive to ESD. Such constructions include microcircuits with metallization paths over N+ guard rings. N+guard rings are used in the N-type epitaxial islands to inhibit possible inversion of the N-type semiconductor to P-type semiconductor and to reduce the leakage current. Since the final oxide layer over the N+ guard ring is relatively thin, parasitic MOS capacitors of relatively low breakdown voltage are created when a metallization path passes over this ring. These MOS capacitor structures are ESDS;

l. Caution is advised in the use of microcircuits and hybrids containing dielectrically isolated bipolar parts which are generally moderately sensitive to ESD. Failure occurs due to breakdown of the thin dielectric layers between these small geometry parts from an ESD.

A.1.2.2 Assembly Design Considerations

Procedures are as follows:

a. Latchup in CMOS, with the exception of analog switches, can be avoided by limiting output current. One solution is to isolate each output from its cable line with a resistor and clamp the lines to V_{DD} and V_{SS} with two high speed switching diodes. The use of long input cables poses the possibility of noise pickup. In such cases filter networks should be used;

b. Additional protection can be obtained from MOS by adding external series resistors to each input;

c. Where practicable, an RC network consisting of a relatively large value resistor and a capacitor of at least 100 pF should be used for sensitive inputs on bipolar parts to reduce effects from ESD. However, if circuit performance dictates, two parallel diodes clamping to a half volt in either polarity can be used to shunt the input to ground. This reduces disturbances to the input characteristics;

d. Leads of sensitive parts mounted on PWBs should not be connected directly to connector terminals without series resistance, shunts, clamps or other protective means. Assembly designs containing ESDS items should be reviewed for incorporation of protective circuitry.

A.1.3 ESD Part Protection Networks

Manufacturers have incorporated protection circuitry on most MOS devices (see Fig. A.1). The purpose of these protection networks is to reduce the voltage across the gate oxide below the dielectric

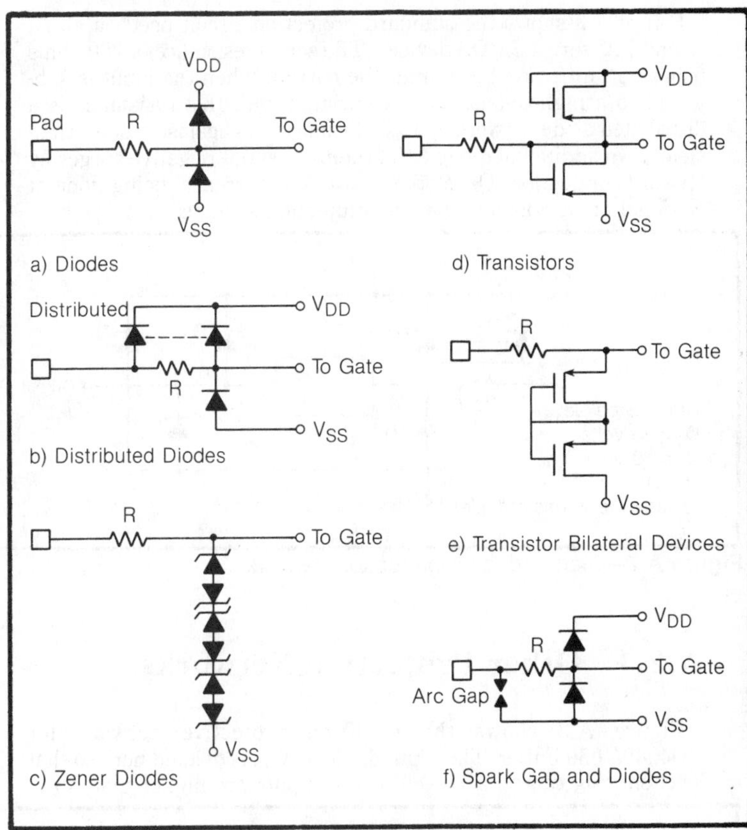

Figure A.1—Gate protection networks

breakdown voltage without interfering with part electrical performance. Differences in fabrication processes, design philosophies and circuitry have resulted in difference gate protection networks.

A.2 Standard Input Protection Networks*

In order to protect the gate oxide against moderate levels of electrostatic discharge, protective networks are provided on all National CMOS devices, as described below.

*Section A.2 is taken from National Semiconductor, a leader in state-of-the-art technology, Application Note 248

Figure A.2 shows the standard protection circuit used on all A, B, and 74C series CMOS devices. The series resistance of 200 ohms using a P^+ diffusion helps limit the current when the input is subjected to a high-voltage zap. Associated with this resistance is a distributed diode network to V_{DD} which protects against positive transients. An additional diode to V_{SS} helps to shunt negative surges by forward conduction. Development work is currently being done at National on various other input protection schemes.

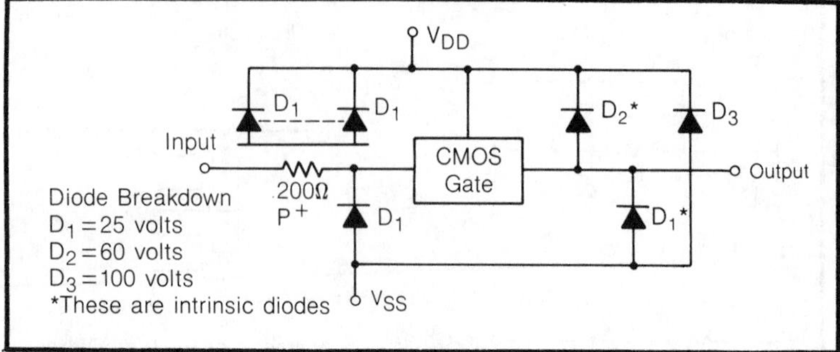

Figure A.2—Standard input protection network

A.2.1 Other Protective Networks

Figure A.3 shows the modified protective network for CD4049/4050 buffer. The input diode to V_{DD} is deleted here so that level shifting can be achieved where inputs are higher than V_{DD}.

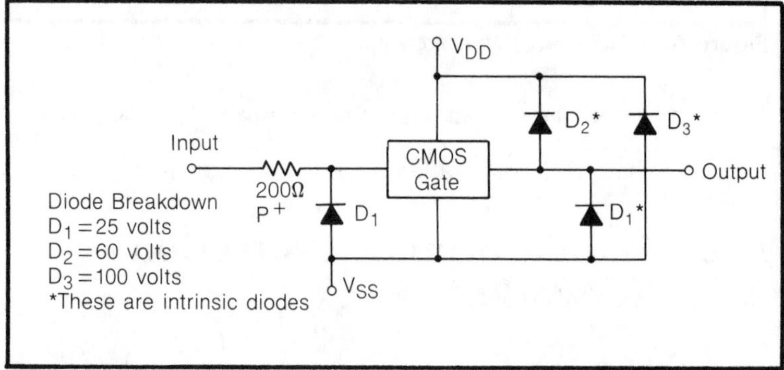

Figure A.3—Protective network for CD4049/50 and MM74C901/2

Figure A.4 shows a transmission gate with the intrinsic diode protection. No additional series resistors are used so the on resistance of the transmission gate is not affected.

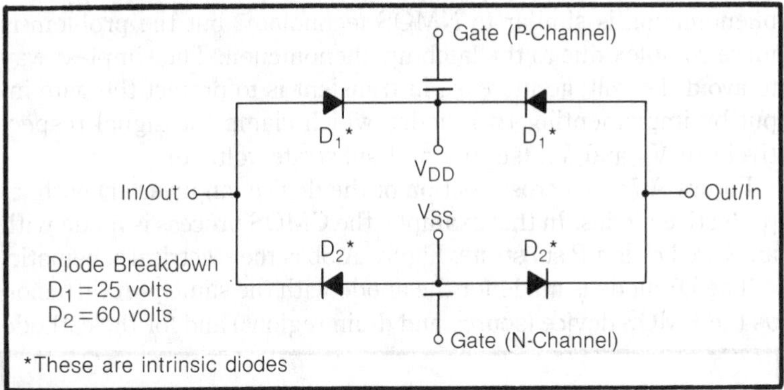

Figure A.4—Transmission gate with intrinsic diodes to protect against static discharge

A.3 Manufacturing Aspects of MOS Overvoltage Protection

A.3.1 The Problem

The gate dielectric material of insulated MOS or SGFeT is very sensitive to breakdown (during fabrication or test) as its thickness decreases to the 250 Å range. High voltage pulses of static electricity can easily destroy the device. Protection techniques are now extensively used, and the circuits must meet the following conditions:

- No extraneous steps introduced in the process (this reduces cost)
- Minimum load attached to the FeT input during usual operation
- Minimum response time to divert the voltage surge
- Minimum area not to degrade the I/O cells density although able to sink or supply high currents

A.3.2 The Diode Input Structure

All the following examples are related to CMOS technology. The phenomenon is similar to NMOS technology, but the problem is more complex due to the latch-up phenomenon. The simplest way to avoid the voltage increase in transient is to protect the gate input by implementing two diodes which clamp the signal respectively to V_H and V_{sx} (supply and substrate voltage).

Figure A.5a is a cross-section of the device (an inverter) with its protective diodes. In this example, the CMOS process is made with an N well over a P substrate. Figure A.5b is the electrical schematic.

The D1 diode is made for the anode with the same port diffusion as the PMOS device (source and drain regions) and for the cathode

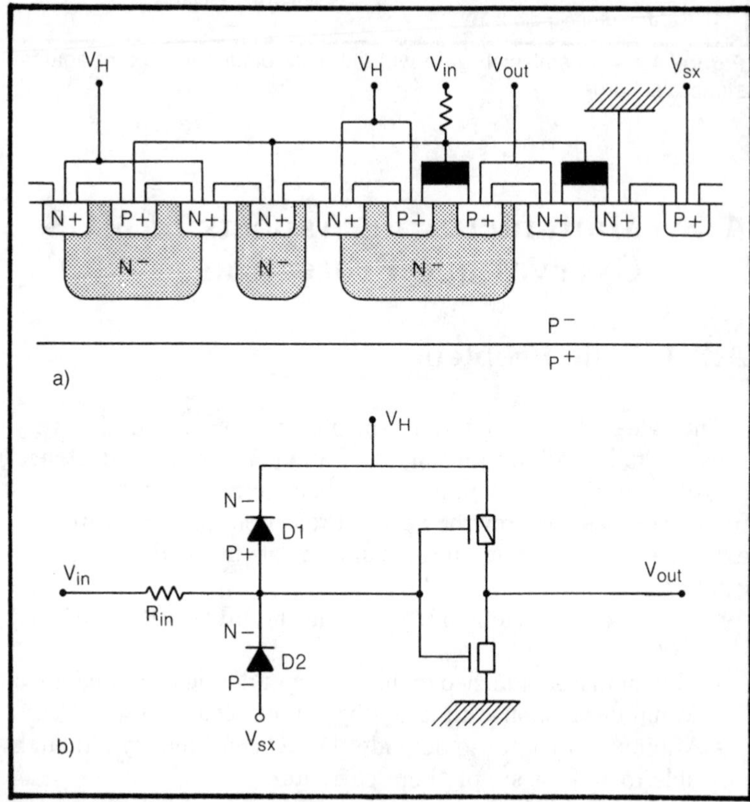

Figure A.5—Cross-section of an inverter with protective diodes

with the N well. Diode D2 is made with the substrate –N well junctions. (Its contact resistance is lowered using an N^+ diffusion—or implantation. In normal operation the input gate is loaded by the two transition capacitances of the reverse diodes.

The current and the energy dissipated in the protective diode can be limited by a resistor in series with the input gate. This can be made from a serpentine of polysilicon (10-50 Ω/square). But if we assume a 100 Ω resistance value associated with an input capacitance of 0.2 pF, the RC product is 0.2 ns. Such a value is not negligible if we consider the present state-of-the-art characterized by unit gate delay in the ns range. Due to the latch-up phenomenon, which will be explained briefly later on, diode D1 is generally suppressed. In this case, the positive discharge is diverted to the substrate (V_{sx}^-) after the D2 diode is turned in reverse condition by avalanche. But this happens at a higher voltage (10-20 V), which in some cases could be closer to the input gate breakdown itself. The reverse breakdown voltage can be lowered just by using the N^+ P^- diode (≅10 V) or by the implementation of a field relief electrode (see Sec. A.3.3).

A.3.3 The Transistor Input Structure

When diode D2 is in avalanche condition, the energy dissipated is very high. (See the typical characteristic in Fig. A.6.) In addi-

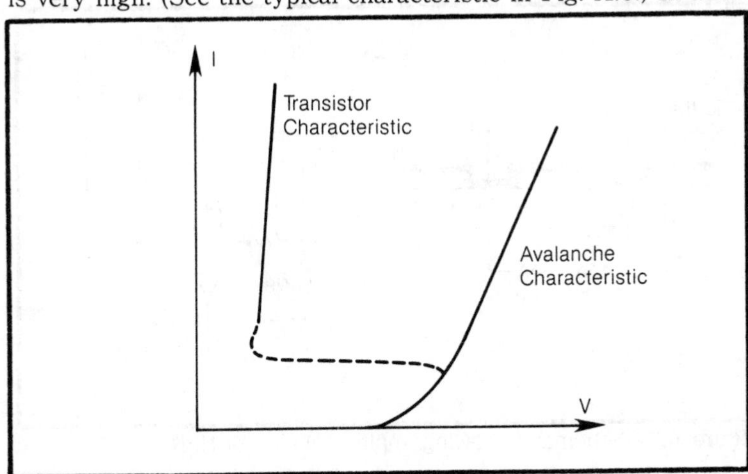

Figure A.6—Typical avalanche condition

tion, the response time is not negligible (function of carrier's concentration). To encompass these severe drawbacks, a new technique using a lateral transistor in diode configuration is presented.

Figure A.7 describes the way the lateral NPN is implemented, as well as showing the electrical schematic.

When the overvoltage occurs, the collector-base junction of the NPN (N^+ P^-) avalanches first. This value can be reduced by an interesting feature of the layout. If the connection of the emitter diffusion is covering the base up to the collector area, the electrical field at the collector-base junction increases (see Fig. A.8); thus, the avalanche breakdown voltage decreases (field relief electrode).

If we change the thickness of the oxide layer over the collector junction, the avalanche point may be totally controlled. But in this case a new step is introduced in the process. The advantage of this technique consists of a fast response time and a reduced resistive path to the substrate. As soon as the avalanche starts, a small current flows in RA, RB, or in the NPN base region. The voltage drop develops until the base emitter junction becomes forward biased. At this time a low impedance path has been initiated. Figure A.6 shows that the increasing current does not follow the avalanche curve, but it abruptly departs from a point where bipolar action is initiated and continues with the classicial transistor characteristic.

It is obvious that the two N^+ diffusions must be placed at a distance in such a way that the transistor operation is feasible and that punch through does not occur. Another parameter has to be

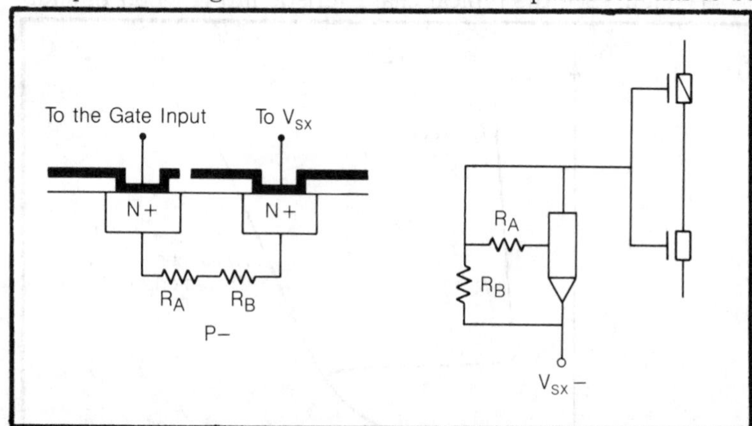

Figure A.7—Schematic showing implementation of NPN

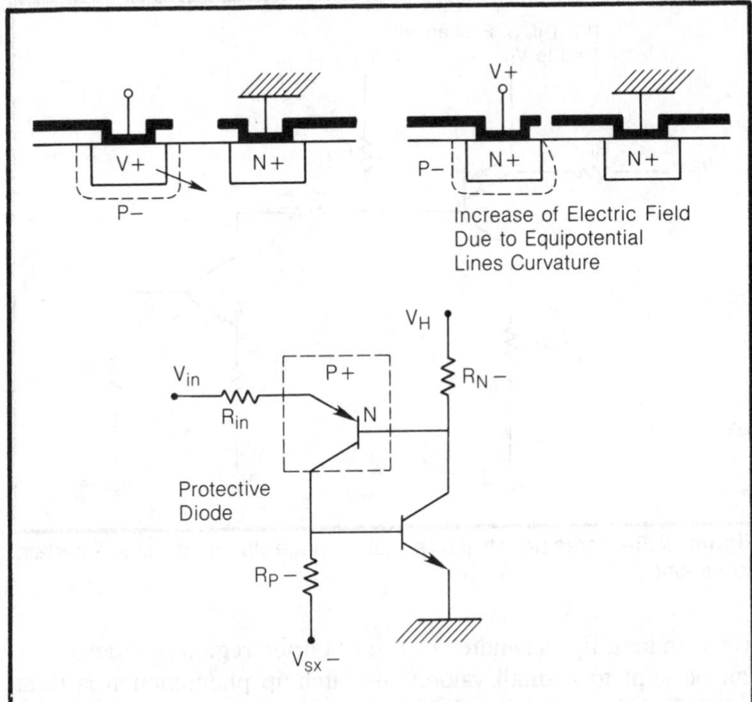

Figure A.8—Parasitic structure

taken into account because the NPN can also be considered a MOS with a grounded, gated diode. Thus, when the overvoltage is not present, some leakage can occur between the diffusions considered as drain-source. This can be related to a high β of the NPN. In summary, a compromise has to be made in the layout between the response time and the leakage.

A.3.4 Protective Diode Drawback

Figure A.9 presents a parasitic structure described in Fig. A.8 which, under certain conditions, can induce latch-up. R_n^- and R_p^- are the resistances of the N well and P substrate regions, respec-

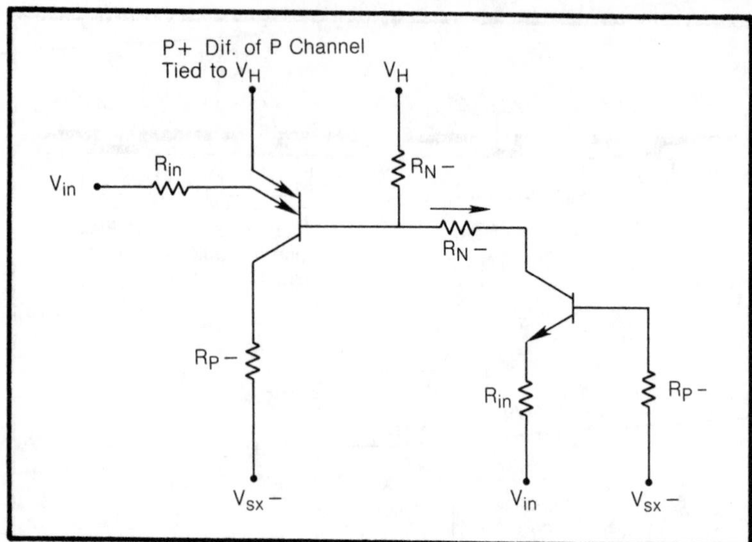

Figure A.9—Parasitic structure can induce latch-up under certain conditions

tively. In fact, R_p^- is shunted by the P^+ buried region resistance and can be kept to a small value. The latch-up phenomenon is thus induced by the protective diode and can be clearly explained. When V_{in} (external to the chip) becomes higher than V_H (overvoltage, parasitic noise), then P^+N junction is forward biased. If the currrent turns in the relation $\alpha PNP + \alpha NPN \geqslant 1$, the SCR is fired. The current is only limited by an external load and generally leads to device destruction. Due to the transistor gains the relation is easily verified. The phenomenon is more difficult if the V_{sx} is tied not to the ground but to a more negative voltage. Even if V_{in} is lowered under the base voltage of the PNP, the P^+ diffusion of a P channel device acts like another emitter and the SCR still remains on. The other clamping diode $N^+N^-P^-$ can also be responsible for latch-up (see Fig. A.9).

When V_{in} is lower than V_{sx}, the NPN is turned on, so a voltage drop occurs as R_n^-. The PNP emitter tied to V_{in} is off, but the one tied to V_H can turn on the PNP as well as the SCR.

Some actions can be taken in the layout to prevent these parasitic actions. Others are related to process parameters (gold doping, epitaxial subtrate, SOS):

- Put the P^+N^- clamping diode far from the N^- well of the active device (to decrease the lateral NPN gain). This also decreases the capacitive coupling between input and internal wiring.
- Decrease resistances of N^- well and P^- substrate by using tapped contacts whenever possible.
- Maximize R_{in}, but this is gated by the circuit performance (and the degradation allowed).
- Put a Ring N^- (Fig. A.10) acting as a collector for the lateral NPN. The ring is tied to the most positive voltage and surrounds the clamp diode. The injected electrons coming from the N^- region of the P channel device are trapped by the ring and do not contribute to the base current of the $P^+N^-P^-$ (clamping diode-substrate).

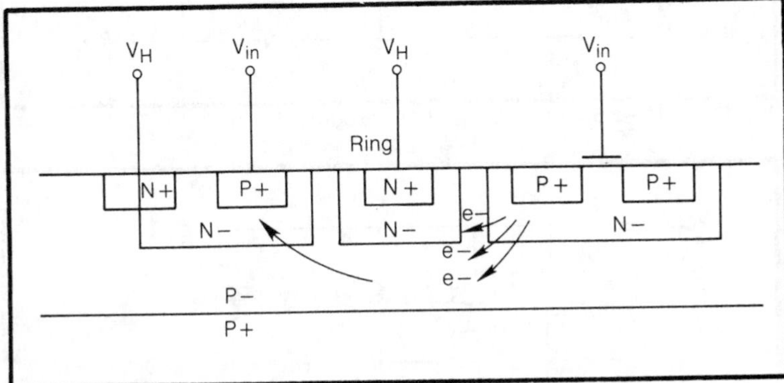

Figure A.10—N^- ring acts a collector for the lateral NPN

A.4 Protection Scheme for High Speed Logics*

A.4.1 HCMOS

TI's unique input protection circuitry provides immunity typically to 4500 V.

*Section A.4.1 is taken from Texas Instruments (TI) Information Journal and Technical Notes

Figure A.11 shows the circuitry implemented to protect gates against ESD. The diode is forward biased for input voltages greater than V_{CC}+0.5 V. The two transistors and resistor (actually one transistor diffused across a resistor) act as a resistor-diode network against negative-going transients. As illustrated in Figure A.12, the ESD protection for the output consists of an additional diffused diode (D3) from the output to V_{CC}. The other two diodes (D1 and D2) are parasitics.

TI's design also incorporates latch-up protection:

Most CMOS devices have two parasitic bipolar transistors: PNP

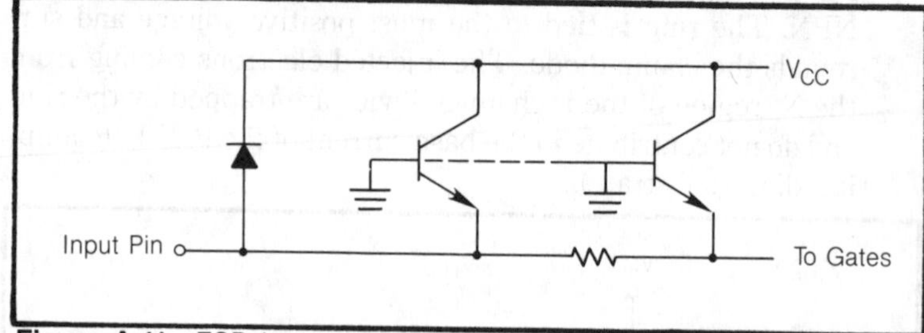

Figure A.11—ESD input protection circuitry

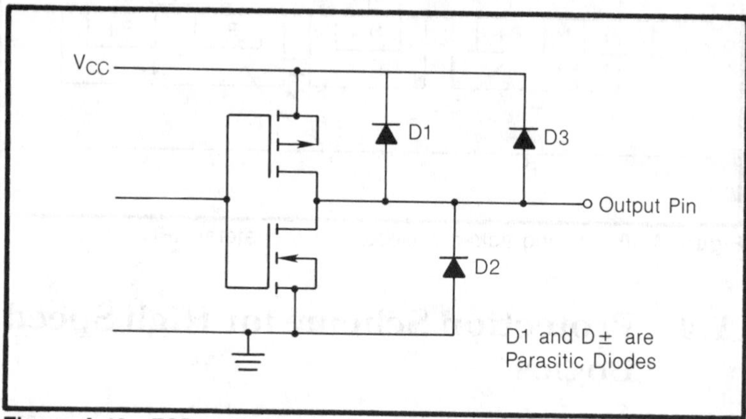

Figure A.12—ESD output protection circuitry

and NPN. Figure A.13 shows the cross-section of a typical CMOS inverter with the parasitic bipolar transistors. Note that, as shown in Fig. A.14, these parasitic transistors are naturally configured as an SCR (Silicon Controlled Rectifier). These transistors conduct when one or more of the PN junctions become forward biased. When this happens each parasitic transistor supplies the necessary base current for the other to remain in saturation. This is known as the "latch-up" condition. This could possibly destroy the device if the supply current is not limited.

A conventional SCR is fired (or turned on) by applying a voltage to the base of the NPN transistor, but the parasitic SCR is fired by applying a voltage to the emitter of either transistor. One emitter of the PNP transistor is connected to an emitter of the NPN transistor which is also the output of the CMOS gate. The other two emitters of the PNP and NPN transistors are connected to V_{cc} and ground, respectively. Therefore, to trigger the SCR, there must be

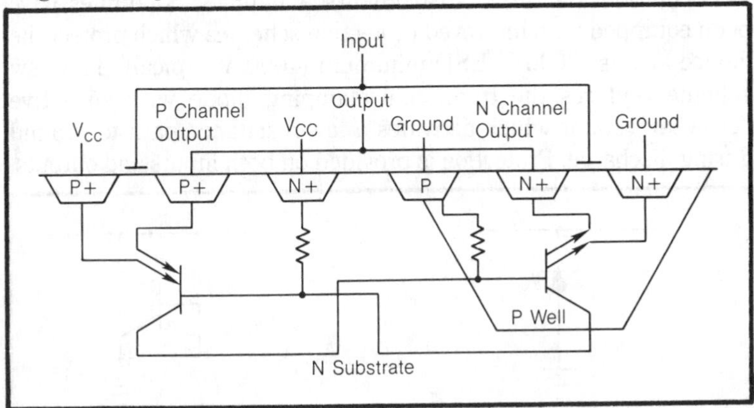

Figure A.13—Parasitic bipolar transistors in CMOS

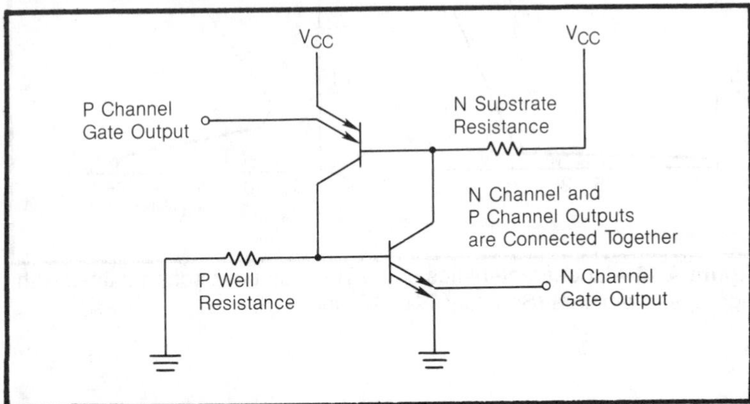

Figure A.14—Schematic of parasitic SCR

a voltage greater than V_{CC}+0.5 V or less than −0.5 V, and there has to be sufficient current to cause the latch-up condition.

Latch-up cannot be completely eliminated. The alternative is to impede the SCR from triggering. Texas Instruments has improved the circuit design by adding four additional diffusions or guardrings alternately connected to V_{CC} and ground.

Four guardrings provide isolation between any IC pin and any PN junction that is not isolated by a transistor gate. All internal PN junctions are separated by two guardrings. Tests have shown effective latch-up resistance ranges from 450 mA to greater than 1 A at 25°C, and typically greater than 250 mA at 125°C.

A.4.2 "Advanced" Schottky

Since 1983, the Texas Instruments AS and ALS families have been equipped with improved protection schemes which protect the device against 2000 V ESD minimum (4000 V typical). The new scheme replaces the traditional clamping diode with an active "crowbar" shunt which activates a low-resistance path to ground during discharge. Protection is provided on both inputs and outputs.

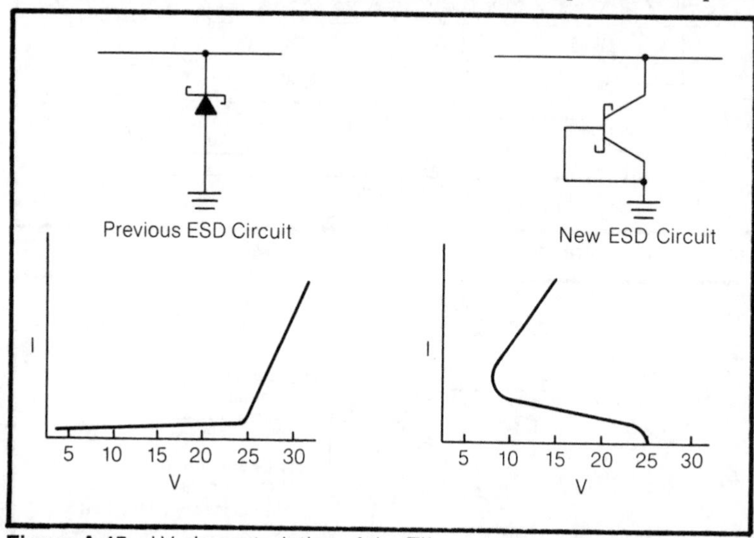

Figure A.15—I,V characteristics of the TI's protective scheme used with high speed bipolars (Schottky TTL, AS and ALS)

Appendix B

Prediction of the ESD Voltage at Which a Given Semiconductor Junction Will be Damaged

Using some of the simple models of device failure given in Ref. 1, the human body voltage at which a semiconductor junction may be damaged can be approximated.

Illustrative Example—Integrated Transistor (Fig. B.1)

The emitter-base junction being the smallest physical area and being connected directly to the package leads, it is assumed that it will be the first candidate for failure. The following constants will

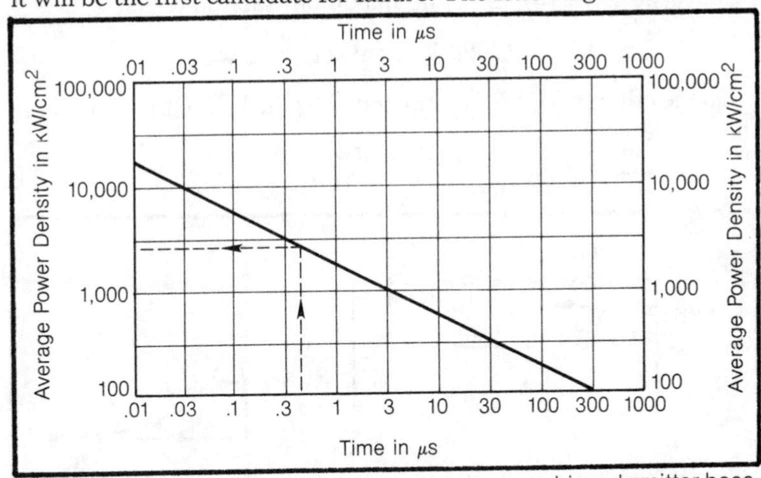

Figure B.1—Average power density across a reverse biased emitter base junction (NPN transistor) causing failure of ⩾5% of devices (Ref. 1)

be used:

- Cross-section offered to the current path=10^{-6} cm²
- R_B, internal resistance near the emitter-base=25 Ω
- V_{RBD}, reverse breakdown voltage=10 V

Human body parameters will be chosen similar to the worst-case assumptions of the human body models (for example, MIL-STD-883 test):

- R_H=1500 Ω
- C_H=100 pF

The discharge time constant of the circuit can be calculated. If we neglect the other contact resistances in series in the path (such as chip-to-leads, leads-to-finger, lead-to-ground, etc.) since they represent less than a few ohms, it becomes:

$$\tau_c=(R_B+R_H)\ C_H$$

$$\tau_c=(25+1500)\ 100\cdot10^{-12}=152\ \text{ns}$$

If we consider that the full pulse width is reached in about $5\times\tau_c$, or 760 ns, we will be able to determine the maximum permissible power averaged over 760 ns.

From Fig. B.1, in the microsecond range, we see that the maximum power density that a reverse biased silicon PN junction can handle is about 2000 kW/cm² for 0.76 μs, corresponding to 1.5 Joule/cm².

Since we have 10^{-6} cm², the maximum safe power is:

$$P_{D\ max}=(2000\ \text{kW/cm}^2)\times10^{-6}=2\ \text{watts}$$

On the other hand, the power created by the ESD current through the device is:

$$P=V_{RBD}i+R_Bi^2 \text{ with } i=I_pe^{-t/\tau}$$

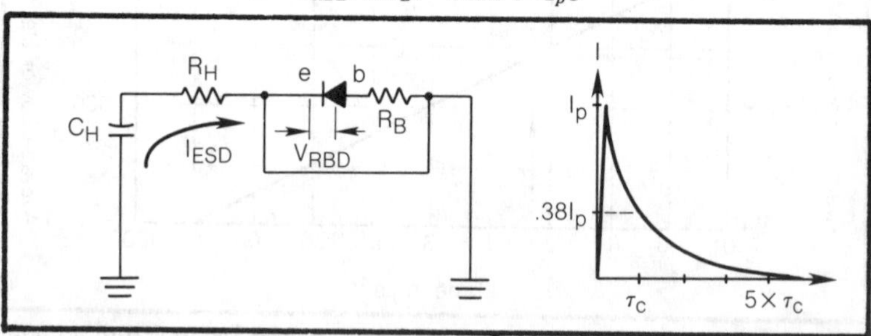

Figure B.2—Simplified equivalent circuit for damage threshold approximation

where the first term represents the power in the reverse biased PN barrier, and the second term represents the power dissipated in the internal device resistance near the emitter-base junction. Replacing i with its value yields:

$$P = V^{RBD}\, I_p e^{-t/\tau} + R_B\, (I_p e^{-t/\tau})^2$$

$$= V_{RBD}\, I_p e^{-t/\tau} + R_B I^2 p e^{-2t/\tau}$$

Averaging the expression over $t = 5\times\tau$, the following is obtained:

$$P_{AV} = \frac{1}{5\tau}\int_0^{5\tau} V_{RBD}\, I_p e^{-t/\tau}\, dt + \frac{1}{5\tau}\int_0^{5\tau} R_B\, I^2 p e^{-2t/\tau}\, dt$$

$$= \frac{V_{RBD}\, I_p}{5}\,(1-e^{-5}) + \frac{R_B I^2{}_p}{10}\,(1-e^{-10})$$

$$\cong \frac{V_{RBD}\, I_p}{5} + \frac{R_B I_p{}^2}{10}$$

$$\text{since } I_p = \frac{V_{ESD} - V_{RBD}}{R_H + R_B} \cong \frac{V_{ESD}}{1500+25}$$

Replacing P_{AV} with the limit calculated before and I_p with its value gives:

$$2 = \frac{10}{5}\;\frac{V_{ESD}}{1525} + \frac{25}{10}\left(\frac{V_{ESD}}{1525}\right)^2$$

Solving for V_{ESD} and neglecting the imaginary root, we get:

$$V_{ESD} = 365 \text{ volts}$$

This integrated circuit is vulnerable if handled by a person charged to >365 V.

A more accurate calculation could be carried out by considering that V_{RBD} varies with current and temperature, both of which vary during the 760 ns discharge duration.

Appendix C

Sparkover Voltages

The voltage at which an air gap will arc depends on the shape of the electrode, the pressure, and the temperature of the air. Provided that the voltage is dc or at a frequency low enough to allow a complete de-ionization of the channel between two consecutive arcs, and provided that the gap itself is dry and dust-free, the sparkover voltages are given by the Paschen curves in Figs. C.1 and C.2.

These curves show that around 1 atmosphere, the breakdown voltage is approximately proportional to pressure and gap length. It is also inversely proportional to absolute temperature. Certain synthetic gases have higher arcing voltages than air. Sulfur hexafluoride (SF_6) and Freon 12 (CCl_2F_2) have a dielectric rigidity about 2.5 times higher than air. This is why such gases, enclosed in a bulb at high pressure, can make high voltage relays with minimal arcing/bouncing problems (see Sec. 4.1).

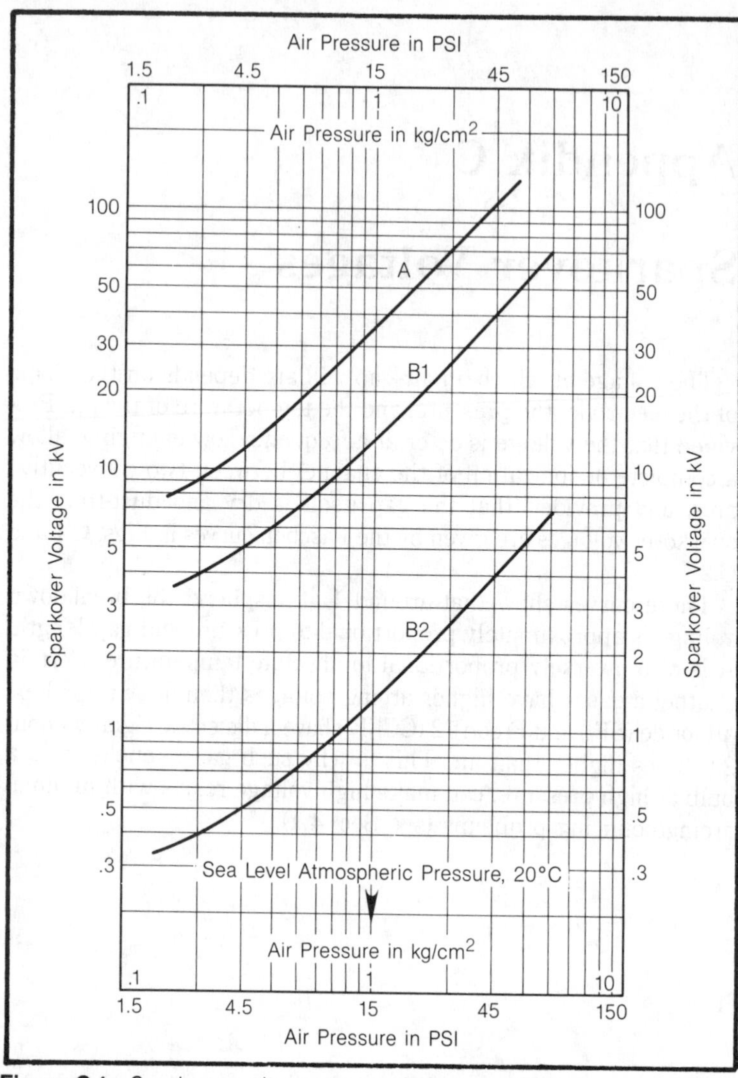

Figure C.1—Sparkover voltage versus air pressure (Paschen curves) for:
A : smooth spherical electrodes with diameter ⩾gap distance, for a 1 cm gap
B1: needle gap, 1 cm
B2: needle gap, 1 mm

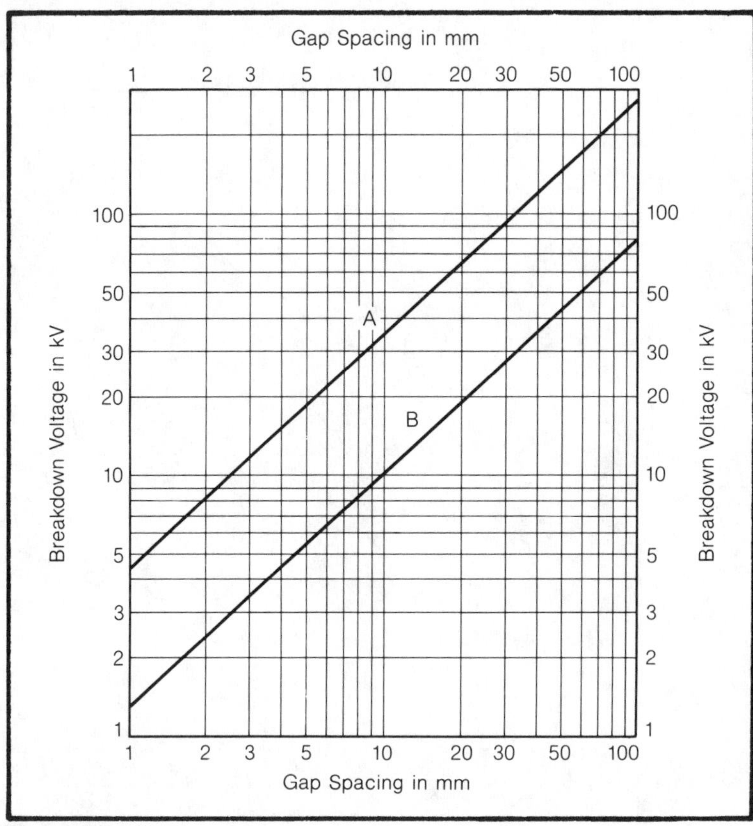

Figure C.2—Sparkover voltage versus gap spacing at normal atmospheric pressure

A: smooth spherical electrodes with diameter≥gap distance

B: needle gap

Appendix D

The Fatigue Phenomenon During Repeated ESD Testing

Chapter 2 addressed the problem of the "walking woundeds"; i.e., devices which still function after repeated ESD pulses, but whose life span has been significantly reduced. All the specifications relative to device testing acknowledge this fact by requesting that all samples which have been submitted for testing not be used for production.

However, what about the hundreds of integrated circuits which are stressed when the complete machine is submitted to ESD testing? One could argue that this is a second degree problem because these parts are not zapped directly and are subjected to much lesser voltages that are the result of induction coupling with low energy. This argument is not totally true.

- A direct ESD can strike a component pin, in some cases, via a connector pin, a keyboard pad, or an exposed LED or switch.
- Even with purely indirect discharge, the repetitive nature of an intensive ESD test can *impose on some exposed modules several thousands of transients* reaching the hundreds of volts range. Keep in mind also that, unlike module testing which is generally done at 20°C, the ambient inside the EUT during a machine test is at 50° or 55°C with the junction temperatures approaching 85°C. This causes at least three aggravating conditions:

  —the breakdown voltages decrease when t° increases

  —all the fatigue phenomena accelerate exponentially with temperature, following the Arrhenius theory

—all the electro-migration phenomena also increase with temperature. They tend to:
- grow ramifications from one conductor to another; they locally increase the field gradient, and the problem becomes self-aggravating
- create aluminum pile-up at some places and thinning at others until current densities of 10^8 or 10^9 A/cm are reached where cracks or melting can occur.

So far, none of these aspects has been considered during equipment testing. It may very well be that this problem is a false one, but it would certainly be worthwhile to consider it to ascertain whether an EUT which has endured several cycles of ESD checks and is then put on the market has a reduced life-span.

Appendix E

Time to Frequency Conversion of a Single Transient

An isolated pulse (voltage, current, E or H-field, etc.) which has no repetition period can be translated into the frequency domain as a broadband spectrum where individual harmonics do not exist, since the Fourier series is being replaced by the Fourier integral. Since the time domain waveform area could be sized in volts$\times$seconds or amperes$\times$seconds, the corresponding spectral density can be expressed in volts/Hz or amperes/Hz with multiples and sub-multiples of μV/MHz or μA/MHz.

	Time Domain	Frequency Domain	Spectrum Equations
Damped Oscillatory (Arc Discharge and Filter Switching Transient)	A, A_1, A_2, T_1, T_o	F1, A_N, 40 dB/dec., Frequency (F_n)	$F_n < 0.5F_1 < \frac{1}{8T_1}$ $A_N = 126 + 20\log_{10} AT_1$
			$F_n = F_1 = \frac{1}{4T_1}$ $A_N = 126 + 20\log_{10} AT_1 - 10\log_{10}[4\delta^2(1-\delta)]$
			$F_n > 2F_1 > \frac{1}{2T_1}$ $A_N = 96 + 20\log_{10}\frac{A}{T_1} - 40\log_{10} F_n$
Over Damped Unidirectional (Arc Discharge and Relay Switching Transient)	A, T_1, T_E, T_o	A_N, F_1, F_2, 20 dB/dec., 40 dB/dec., Frequency (F_n)	$F_n < F_1 < \frac{1}{\pi T_E}$ $A_N = 120 + 20\log_{10} AT_E$
			$F_n > F_1 > \frac{1}{\pi T_E}$ $A_N = 110 + 20\log_{10} A - 20\log_{10} F_n$
			$F_n > F_2 > \frac{1}{\pi T_1}$ $A_N = 100 + 20\log_{10}\frac{A}{\pi T_1} - 40\log_{10} F_n$

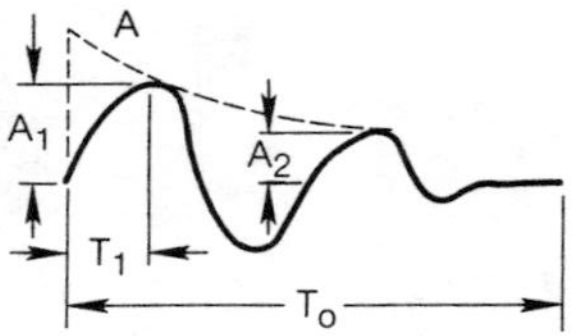

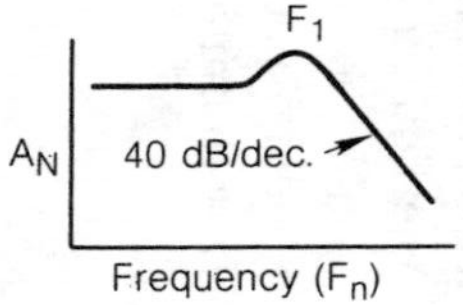

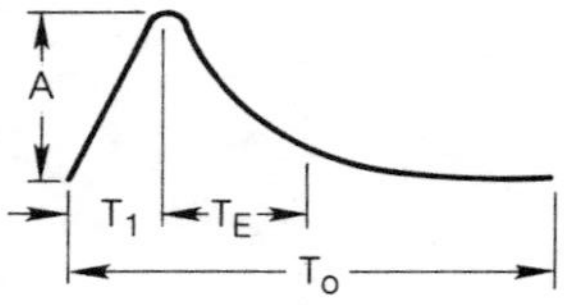

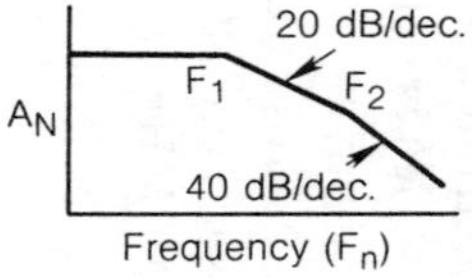

A = Volts or Amperes (peak)
T = Time in microseconds
A_N = dBμV/MHz or dBμA/MHz
F = Frequency in MHz
π = 3.1416
R = Resistance in Ω
L = Inductance in μH
dec. = decade
C = Capacitive in microfarads

*Damping ratio $(\delta) = \frac{\ell_n A_1 - \ell_n A_2}{2\pi} = \frac{R}{2}\sqrt{C/L}$

Note: T_E is the time for A to drop to a value of 1/8 A

References

1a. Jowett, C.E. *Electrostatics in the Electronics Environment.* United Kingdom: Halsted Press Book/New York: John Wiley & Sons.

1b. Horvath, A. and Berta. *Static Elimination.* United Kingdom: Research Studies Press/New York: John Wiley & Sons.

2. Testone, A. "Static Electricity in Electronics Industry." Testone Enterprises, Box 323, Lee, Massachusetts.

3. Aguet, M. "Decharges d'Origine electrostatique." Bulletin SEV/VSE 74, Switzerland, 1983.

4. Richman, P. *ESD Protection Handbook.* KeyTek Instrument Corp.

5. Kirk, W. "Eliminate Static Discharge." *Electronic Design,* March 1976.

6. Yenni, D., J. Huntsman and G. Mueller. 3M

6b. Petrizio. "Electrical Overstress vs. Device Geometry." Electical Overstress/ESD Symposium, Denver, Colorado, 1979.

7. King, M. "Dynamic Waveforms of Personnel ESD." Electrical Overstress/ESD Symposium, Denver, Colorado, 1979.

8. King. M. "Impulse Waveforms of Personnel/Furniture ESD." IEEE/EMC Symposium, Santa Clara, California, 1982.

9. Madzy, T. "Static Discharge Modelling." EMC Symposium, Montreux, Switzerland, 1975.

10. Simonic, R. "Personnel ESD Statistics." IEEE/EMC Symposium, Boulder, Colorado, 1981.

11. Simonic, R. "ESD Furniture Event Rates." IEEE/EMC Symposium, Santa Clara, California, 1982.

12. Hyatt, H. and H. Mellberg. "Bringing ESD Testing into the 20th Century." IEEE/EMC Symposium, Santa Clara, California, 1982.

13. Byrne, W. "Development of an ESD Model for Electronic Systems." IEEE/EMC Symposium, Santa Clara, California, 1982.

14. Tucker, T.J. "Spark Initiation Requirements." *Annals of the New York Academy of Sciences,* Vol. 152, Oct. 28, 1968, Art. 1.
15. Ryser, H. and B. Daout. "Fast Discharge Mode in ESD." Zurich EMC Symposium, 1985.
16. Richman, P. "ESD Testing: The Interference Between Simulator and E.U.T." Zurich EMC Symposium, 1985.
17. King, Michael. Cornell Dubilier Report on EMI Susceptibility. December 1983.
18. Chase, E.W. "ESD Susceptibility of Thin Film Resistors." EOS Symposium, Orlando, Florida, 1982.
19. Bossard and Unger. "ESD Damage from Charged IC Pins." EOS Symposium, San Diego, California, 1980.
20. Denson and Dey. "ESD Testing of Advanced Schottky TTL." EOS Symposium, Orlando, Florida, 1982.
21. White, Donald R.J. and Michel Mardiguian. *EMI Control Methodology and Procedures.* Fourth Edition. Gainesville, Virginia: Interference Control Technologies, Inc. 1985.
22. Kendall, C. and E. Black. "A General ESD Testing Regimen." IEEE/EMC Symposium, Santa Clara, California, 1982.
23. Branberg, G. "ESD and CMOS Logic." EOS Symposium, Denver, Colorado, 1979.
24. IEC 801-2.
25. Pratt, D. and J. Davis. "ESD Failure Rate Prediction." IEEE/EMC Symposium, 1984.
26. Richman, P. "A Realistic ESD Test Program." *EMC Technology Magazine,* July 1983.
27. Vrachnas, S. "Testing Switches for ESD Simulators." *Electronic Test,* February 1985.
28. Shaw and Enoch. "A Programmable Equipment for ESD Testing to Human Body Model." EOS Symposium, 1983.
29. White, Donald R.J. *Electromagnetic Shielding Materials and Performance.* Gainesville, Virginia: Don White Consultants, Inc., 1980.
30. Faught, A. "Shield Evaluation Using Transfer Impedance Techniques." IEEE/EMC Symposium, Santa Clara, California, 1982.
31. Palmgren, C. "Shielded Flat Cables for EMI and ESD Reduction." IEEE/EMC Symposium, 1981 and *EMC Technology Magazine,* July 1982.
32. Simon, R. and D. Stutz. "Test Methods for Shielding Materials." *EMC Technology Magazine,* October 1983.
33. Vance, E. *Coupling to Shielded Cables.* New York: John Wiley & Sons.

Index

A

B

C

G

H

I

J

Q

R

S

T

U

V

W

Z